28th Joint Seminar Development of Materials Science in Research and Education (DMSRE28)

IOP Conference Series: Materials Science and Engineering
Volume 465

Pavlov, Czech Republic
3 – 7 September 2018

ISBN: 978-1-5108-8095-5
ISSN: 1757-8981

Printed from e-media with permission by:

Curran Associates, Inc.
57 Morehouse Lane
Red Hook, NY 12571

Some format issues inherent in the e-media version may also appear in this print version.

Printed by Curran Associates, Inc. (2019)

Institute of Physics
Dirac House, Temple Back
Bristol BS1 6BE UK

Phone: 44 1 17 929 7481
Fax: 44 1 17 920 0979

techtracking@iop.org

Additional copies of this publication are available from:

Curran Associates, Inc.
57 Morehouse Lane
Red Hook, NY 12571 USA
Phone: 845-758-0400
Fax: 845-758-2633
Email: curran@proceedings.com
Web: www.proceedings.com

Table of contents

Volume 465

Development of Materials Science in Research and Education (DMSRE28)

3–7 September 2018, Pavlov, Czech Republic

Accepted papers received: 16 November 2018
Published online: 4 January 2019

Preface

Papers

Preface

The 28th seminar "Development of Materials Science in Research and Education" held on 3-7 September 2018 in hotel IRIS Pavlov, Czech Republic. The first Joint Seminar in these series was held at Gabčíkovo in 1991 by the initiative of the Czech and Slovak Association for Crystal Growth and the Slovak Expert Group of Solid State Chemistry and Physics. The objective of this meeting offered an opportunity to Czech and Slovak teachers and scientists as well as guests from other countries who are working in the field of Materials Science to present their recent results and experience and to exchange new ideas and information.

The scientific sessions covered the following topics on materials science:

- Trends in the development of materials research
- Education of materials science at the universities
- Information about the research programmes of individual institutions
- Information about equipment for preparation and characterization of materials
- Results of materials research

This workshop was aimed at the creation of a stimulating atmosphere of cooperation and at the support of patient dissemination of scientific ideas and propagation of materials science in education. At the seminar, 48 participants presented 43 interesting talks.

Editors

28th Joint Seminar

Development of Materials Science in Research and Education

Hotel IRIS Pavlov, 3 - 7 September 2018

Organized by

Czechoslovak Association for Crystal Growth
Slovak Expert Group of Solid State Chemistry and Physics

Under the auspices of

Institute of Physics, Czech Academy of Sciences
Slovak Society for Industrial Chemistry

Scientific Committee

Mária Behúlová	Slovak University of Technology in Bratislava, Slovakia
Jaromír Drápala	Technical University of Ostrava, Ostrava, Czech Republic
Jan Grym	Institute of Photonics and Electronics, Prague, Czech Republic
Marian Koman	Slovak University of Technology in Bratislava, Bratislava, Slovakia
Zdeněk Kožíšek (*chairman*)	Institute of Physics, Czech Academy of Sciences, Prague, Czech Republic
Jiří Luňáček	Technical University of Ostrava, Ostrava, Czech Republic
Maroš Martinkovič	Faculty of Material Science and Technology SUT, Trnava, Slovakia
Petr Mošner	University of Pardubice, Pardubice, Czech Republic
Blažena Papánková	Slovak University of Technology in Bratislava, Bratislava, Slovakia
Jan Polák	Crytur Turnov, Turnov, Czech Republic
Zdeněk Potůček	Faculty of Nuclear Sciences and Physical Engineering CTU, Prague, Czech Republic
Kateřina Rubešová	University of Chemistry and Technology, Prague, Czech Republic

Organizing Committee

Aleš Bystřický	Institute of Physics, Czech Academy of Sciences, Prague, Czech Republic
Zdeněk Kožíšek	Institute of Physics, Czech Academy of Sciences, Prague, Czech Republic
Robert Král	Institute of Physics, Czech Academy of Sciences, Prague, Czech Republic
Petra Zemenová	Institute of Physics, Czech Academy of Sciences, Prague, Czech Republic

Peer review statement

All papers published in this volume of *IOP Conference Series: Materials Science and Engineering* have been peer reviewed through processes administered by the proceedings Editors. Reviews were conducted by expert referees to the professional and scientific standards expected of a proceedings journal published by IOP Publishing.

Induction brazing of thin-walled pipes from AISI 304 steel using copper-based solder

M Behúlová, Š Vrtiel and M Nagy

Slovak University of Technology in Bratislava, Faculty of Materials Science and Technology in Trnava, Ulica Jána Bottu č. 2781/25, 917 24 Trnava, Slovakia

maria.behulova@stuba.sk, stefan.vrtiel@stuba.sk, mate.nagy@stuba.sk

Abstract. Pipes made of austenitic stainless steels are widely used in automotive industry. They are connected together or to other components by different joining technologies. The quality of produced joints is often critically important for passenger safety assurance. The joint lifetime is significantly influenced by their corrosion behaviour. The paper deals with induction brazing of thin-walled pipes from austenitic stainless steel using high corrosion-resistant Cu-Ni solder. To design appropriate parameters of induction heating, the effective method of numerical simulation of induction heating using the software ANSYS was applied. Simulation model for evaluation of electro-magnetic and temperature fields in the process of induction brazing of exhaust pipes from AISI 304 steel using copper-based solder was developed and verified by experimental temperature measurement during pipe induction heating. Based on performed numerical experiments and results of numerical simulations, appropriate parameters and operating diagram for the induction brazing of thin-walled pipes of the AISI 304 steel using copper-based solder were suggested.

1. Introduction

Induction brazing is used in various applications in the automotive, aviation, aerospace, electro-technical or medical industries. A number of materials can be joined using this method including aluminum, brass, copper, copper alloys, iron, steels or stainless steels. Induction brazing is very suitable also for production of dissimilar joints when materials to be joined have quite different material properties [1-3].

Induction brazing provides many benefits comparing with other joining technologies. Induction heating represents a fast, controllable, reproducible and clean method of material heating. Moreover, it is very efficient, energy saving, safe, adaptable for automated processes, and thus compatible with the growing demands of mass production [4-6]. Since the material is heated rapidly in a small localized area, the rest of the components being joined is thermally unaffected what is advantageous for eliminating distortion, stresses, undesirable metallurgical changes, surface corrosion, and oxidation of the components [9-13].

The oxidation and corrosion is one of key problems of butt weld joints in the exhaust system of cars, prepared by fusion welding technologies [14-16]. In some cases, the replacement of butt weld joints by lap brazed joints of exhaust pipes can provide an outstanding solution, mainly when joining the pipes of austenitic stainless steels using proper types of corrosion resistant solders. According to accomplished research [17], the copper-based solder containing from 3 to 10 wt.% of Ni can be a promising choice.

The main aim of this paper is to design the parameters of induction heating for induction brazing of thin-walled pipes from AISI 304 steel by copper-based solder using numerical simulation of the joining process in the ANSYS program code. For this purpose, methodology of numerical simulation of a pipe induction heating was suggested and verified by experiment. Afterwards, simulation model for induction brazing of pipes from AISI 304 steel was developed and applied to study the electromagnetic and temperature fields in brazed pipes and solder during the process.

2. Problem description

Two pipes made of AISI 304 steel with the outer diameters of D_1 = 31.8 mm and D_2 = 33.6 mm, respectively and the same wall thickness of 0.9 mm should be brazed by B-Cu97Ni(B) solder using internal induction heating, i. e. with inductor coil placed inside tubes being brazed. The solder is supplied in the form of rings prepared from a wire of 1 mm in diameter. As inductor, a copper water cooled induction coil with two turns is suggested. The loop diameter is 28 mm, wire outer diameter is 4 mm and the wall thickness is 1 mm. The allowed maximum duration of induction heating applying the frequency of 20 kHz is 12 seconds.

Chemical composition of AISI 304 austenitic stainless steel is given in Table 1. The copper-based EN 1044 B-Cu97Ni(B) solder with the content of 87 wt.% of copper and 0.02 –0.05 wt.% of boron is used for the high-temperature brazing of alloyed and unalloyed steel as well as hard metals, wolfram, molybdenum and tantalum [18-19]. The addition of nickel ensures a higher tensile strength compared to pure copper and better wetting properties of the solder.

Table 1. Chemical composition of AISI 304 steel [20].

Element	C	Cr	Ni	Mn	Si	P	S	Fe
wt. [%]	max. 0.08	18 - 20	8 - 10.5	max. 2.0	max. 1.0	0.045	0.03	bal.

The solidus and liquidus temperatures of the AISI 304 steel are T_S = 1360 °C and T_L = 1440 °C, respectively. The melting point of the applied B-Cu97Ni(B) solder is lower, the solidus temperature T_S = 1080 °C and the liquidus temperature T_L = 1105 °C. The solder working temperature is from the range of 1115 °C to 1125 °C. The main issue is if the required solder working temperature may be achieved using induction heating without melting thin-walled pipes from AISI 304 steel, i. e. if the pipe temperature does not exceed the stainless steel solidus temperature of 1360 °C.

3. Mathematical model

In induction heating processes, an alternating voltage applied to an induction coil terminals produces an alternating electrical current flow in the coil circuit, generating a time-variable magnetic field in the coil surroundings. In the workpiece placed into the magnetic field, eddy currents of the same frequency and the opposite direction are induced and the workpiece material is heated up due to resistive and hysteresis losses. The induced eddy currents and consequently the generated heat are not uniformly distributed throughout the workpiece. Approximately 63 % of induced current and 86 % of generated heat is concentrated in the surface layer of the heated workpiece called penetration or skin depth. At the distance of one penetration depth from the workpiece surface toward its core, the current density j decreases exponentially from the maximum value of j_0 at the surface to the value $j = j_0/e$ [6]. The penetration (skin) depth δ depends on electrical and magnetic properties of material heated and mainly on the frequency of current in the inductor coil according to the relationship [6]

$$\delta = \sqrt{\frac{\rho_{el}}{\pi \mu_0 \mu_r f}} \tag{1}$$

where ρ_{el} is the electrical resistivity of a workpiece material, μ_0 is the permeability of vacuum ($\mu_0 = 4\pi \times 10^{-7}$ H.m^{-1}), μ_r is the relative permeability and f is the frequency of the source current.

Numerical simulation of induction heating processes requires to solve coupled electro-magnetic and transient thermal problem because of temperature dependent electrical and magnetic material properties, time, frequency and material dependent loading. The material properties in both, electro-magnetic and thermal analyses, vary very strongly with temperature. In addition, the boundary conditions may be temperature and/or time dependent as well.

3.1. Governing equations for the electro-magnetic field

The governing equations for the electro-magnetic analysis can be derived from the classical Maxwell´s equations [6, 21]

$$\mathrm{rot}\,\boldsymbol{H} = j + \frac{\partial \boldsymbol{D}}{\partial t} \tag{2}$$

$$\mathrm{rot}\,\boldsymbol{E} = -\frac{\partial \boldsymbol{B}}{\partial t} \tag{3}$$

$$\mathrm{div}\,\boldsymbol{D} = \rho_e \tag{4}$$

$$\mathrm{div}\,\boldsymbol{B} = 0 \tag{5}$$

where $\boldsymbol{D}$ is the vector of electric flux density, $\boldsymbol{E}$ is the electric field intensity, $\boldsymbol{B}$ is the magnetic flux density, $\boldsymbol{H}$ is the magnetic field intensity, ρ_e is the free charge density, j the current density and t is the time.

The field vectors $\boldsymbol{D}$ and $\boldsymbol{E}$ and also $\boldsymbol{B}$ and $\boldsymbol{H}$ are related by material properties through the functions referred as the constitutive equations

$$\boldsymbol{D} = \varepsilon\,\boldsymbol{E} \tag{6}$$

$$\boldsymbol{B} = \mu\,\boldsymbol{H} \tag{7}$$

$$j = \sigma\,\boldsymbol{E} \tag{8}$$

in which σ is the electrical conductivity, ε is the permitivity given by the product of the vacuum permitivity ε_0 and the relative permitivity ε_r ($\varepsilon = \varepsilon_0\,\varepsilon_r$) and μ is the magnetic permeability ($\mu = \mu_0\,\mu_r$).

Since $\boldsymbol{B}$ satisfies a zero divergence condition, $\mathrm{div}\,\boldsymbol{B} = 0$, the magnetic vector potential $\boldsymbol{A}$ can be defined by the relationship

$$\boldsymbol{B} = \mathrm{rot}\,\boldsymbol{A} \tag{9}$$

Then the Maxwell´s equations (2)-(5) lead to following expressions of the diffusion equation for electrically conductive (10) and non-conductive materials (11), respectively

$$\Delta \boldsymbol{A} - \varepsilon \frac{\partial^2 \boldsymbol{A}}{\partial t^2} - \mu\sigma \frac{\partial \boldsymbol{A}}{\partial t} = 0 \tag{10}$$

$$\Delta \boldsymbol{A} - \mu\varepsilon \frac{\partial^2 \boldsymbol{A}}{\partial t^2} = 0 \tag{11}$$

For time harmonic electro-magnetic fields, the current density and magnetic vector potential vary with time according to the relationships

$$j = j_0\, e^{i\omega t} \tag{12}$$

$$\boldsymbol{A} = \boldsymbol{A}_0\, e^{i\omega t} \tag{13}$$

where ω is the angular frequency.

For axisymmetric condition, the magnetic field intensity $\boldsymbol{H}$ acts in the axial direction, the magnetic vector potential $\boldsymbol{A}$ and eddy current have only one component in peripheral direction.

Using Eqs. (12) - (13) and axisymmetric conditions, the Eqs. (10) and (11) can be rewritten to the form

$$\frac{\partial^2 A}{\partial r^2}+\frac{1}{r}\frac{\partial A}{\partial r}-\omega\sigma\mu A=0 \tag{14}$$

$$\frac{\partial^2 A}{\partial r^2}+\frac{1}{r}\frac{\partial A}{\partial r}+\omega^2\varepsilon\mu A=0 \tag{15}$$

3.2. Governing equations for the temperature field

The temperature fields are governed by the Fourier-Kirchhoff´s heat diffusion equation [22] representing the energy conservation law for heat conduction which can be written in the form

$$\rho c_{\mathrm{p}}\frac{\partial T}{\partial t}=\frac{\partial}{\partial x}\left[\lambda_x(T)\frac{\partial T}{\partial x}\right]+\frac{\partial}{\partial y}\left[\lambda_y(T)\frac{\partial T}{\partial y}\right]+\frac{\partial}{\partial z}\left[\lambda_z(T)\frac{\partial T}{\partial z}\right]+\dot{q}_v \tag{16}$$

where T is the temperature, ρ is the density, c_{p} is the specific heat, λ_x, λ_y and λ_z are the thermal conductivities in x, y and z directions, respectively and q_v is the volumetric heat source density, i. e. the heat generated per unit time in a unit volume. In induction heating, this term depends on the Joule heat generation in a heated material due to the eddy currents.

For the heat conduction in an isotropic material ($\lambda=\lambda_x=\lambda_y=\lambda_z$) at axisymmetric conditions, Eq. (16) can be rewritten as follows

$$\rho c_{\mathrm{p}}\frac{\partial T}{\partial t}=\lambda\left(\frac{\partial^2 T}{\partial r^2}+\frac{1}{r}\frac{\partial T}{\partial r}+\frac{\partial^2 T}{\partial z^2}\right)+q_v \tag{17}$$

Usually, three kinds of boundary conditions for heat diffusion equation (16) are used. The boundary condition of the first kind is determined by the defined surface temperature as a function of position, $\boldsymbol{r}$, and time, t

$$T = f_1(\boldsymbol{r}, t) \quad \text{at the surface } S_1 \tag{18}$$

The boundary condition of the second kind expresses the surface conductive heat flux as a function of position and time

$$\boldsymbol{q} = -\lambda\frac{\partial T}{\partial \boldsymbol{n}} = f_2(\boldsymbol{r}, t) \quad \text{at the surface } S_2 \tag{19}$$

where $\partial T/\partial \boldsymbol{n}$ denotes differentiation along the outward-drawn normal $\boldsymbol{n}$ at the boundary surface S_2. This condition is applied as well as at the adiabatic (insulated) surfaces or as a symmetry condition when the surface heat flux is zero.

The most frequently, a boundary condition of the third kind is used. According to this condition, the surface conduction heat flux is equal to the convection heat flux

$$-\lambda\frac{\partial T}{\partial \boldsymbol{n}}=h\left(T_{\mathrm{w}}-T_{\mathrm{f}}\right) \quad \text{at the surface } S_3 \tag{20}$$

where h is the heat transfer coefficient, T_{w} is the workpiece surface temperature and T_{f} is the fluid (surrounding) temperature. Usually, the heat transfer coefficient includes the influence of convection and radiation, so it is given by the sum of convection heat transfer coefficient h_{C} and radiation heat transfer coefficient h_{R}

$$h=h_{\mathrm{C}}+h_{\mathrm{R}}=h_{\mathrm{C}}+\frac{\varepsilon_{\mathrm{e}}\sigma_0\left(T_{\mathrm{w}}^4-T_{\mathrm{f}}^4\right)}{T_{\mathrm{w}}-T_{\mathrm{f}}} \tag{21}$$

where ε_{e} is the emissivity and σ_0 is the Stefan-Boltzmann constant.

3.3. Solution procedure

To solve the problem of induction heating for the process of induction brazing of thin-walled pipes, the sequential field-coupling method [21, 23] provided by the ANSYS program code was applied. A special user subroutine was developed to control the field coupling.

According to Figure 1, the harmonic electro-magnetic analysis was performed in the first step to compute the Joule heat generated owing to the eddy currents flowing in a heated workpiece at the initial workpiece temperature. In the next step, the computed Joule heat was applied as a loading for the transient thermal analysis in order to predict the time-dependent temperature distribution. As the temperature field influences the electro-magnetic field by means of the temperature dependent electro-magnetic properties, the computed temperatures were entered as a load for the electro-magnetic analysis. The procedure was repeated to the final time of induction heating.

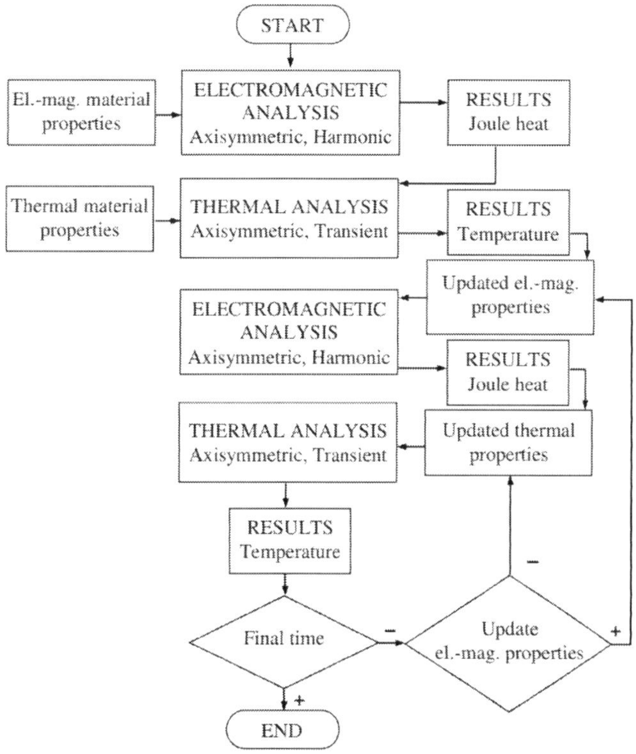

Figure 1. Scheme of solution procedure using the sequential field coupling in ANSYS code.

4. Verification experiment

To verify the described methodology for numerical simulation of induction heating processes, a simple experiment was carried out. The temperatures measured during verification experiment were compared with the results of performed numerical simulation.

4.1. Description of verification experiment

A pipe with the outer diameter of 31.8 mm, the wall thickness of 0.9 mm and the length of 150 mm made of AISI 304 steel (Figure 2) was heated using the two-turn induction copper coil with the same dimensions as inductor proposed for induction brazing, i. e. with the loop diameter of 28 mm, the wire outer diameter of 4 mm and the wall thickness of 1 mm. The position of the induction coil is shown in Figure 3. The parameters of induction heating including applied frequency f, current density in induction coil j_s and heating time t_i are given in Table 2. During experiments, the temperature was measured by two thermocouples located at the outer pipe surface at the distances of Y_1 and Y_2 according to Figure 2 and Table 2.

Table 2. Parameters of induction heating during verification experiment.

Experiment No.	j_s [A.m^{-2}]	f [kHz]	t_i [s]	T_f [°C]	Y_1 [mm]	Y_2 [mm]
1	1.86×10^8	20	12	24.0	37.5	45.5
2	1.78×10^8	20	14.5	25.5	36.5	44.0

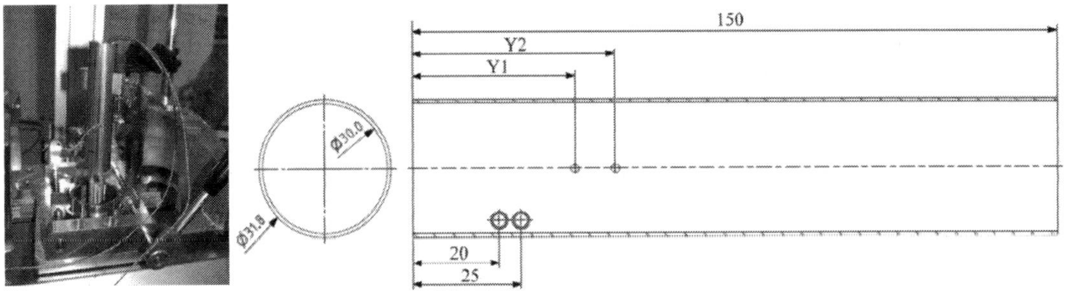

Figure 2. Dimensions of a heated pipe, inductor location and thermocouple positions during verification experiment.

4.2. Simulation model of experimental induction heating

In compliance with Figure 2, geometrical model for numerical simulation of induction heating of a pipe of AISI 304 steel was prepared. The model for electro-magnetic analysis consists of a heated pipe, water cooled induction coil and surrounding air (Figure 3a). In thermal analysis, only the axisymmetric model of a pipe was taken into account. The finite element model (Figure 3b) was generated using the PLANE13 element with the axisymmetric element behaviour option considering the skin depth. The mesh density was higher near the heated pipe surface as well as in the whole computation domain near the induction coil. A detail of the finite element mesh in the region around the inductor is shown in Figure 3c. In the thermal analysis, the element type was switched to PLANE55.

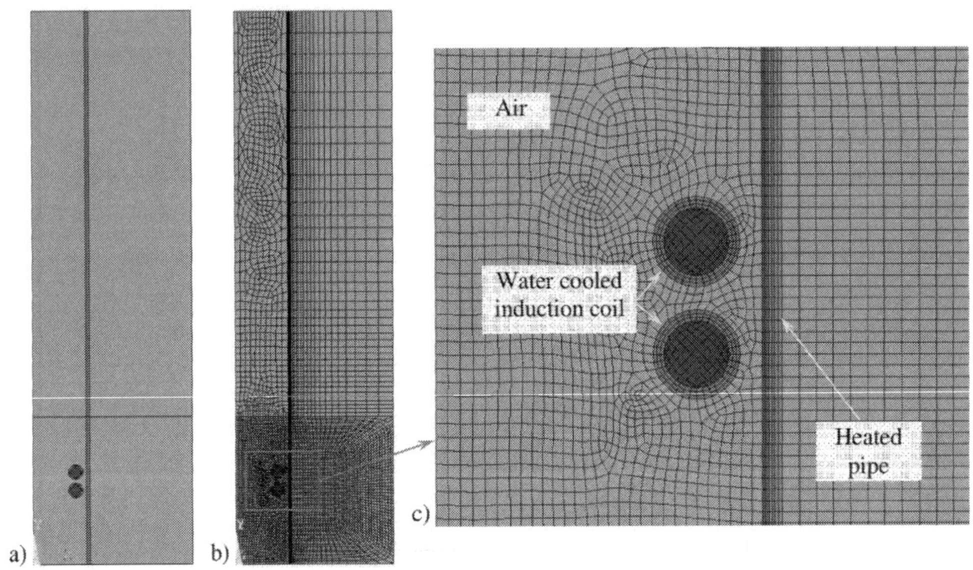

Figure 3. Geometrical model (a) and finite element model (b) for experimental induction heating with a detail of generated mesh near the induction coil (c).

Thermal, electrical and magnetic properties of the AISI 304 austenitic steel in the dependence on temperature were computed using the JMatPro software [24] on the base of chemical composition of the steel presented in Table 1.

Computed values of the thermal conductivity, density, specific heat and electrical resistivity of the AISI 304 steel *vs.* temperature are plotted in Figure 4. The relative permeability and emisivity of the austenitic stainless steel were considered as constant with the values of 1.05 and 0.53, respectively. The solidus and liquidus temperatures of this steel were computed to be 1360 °C and 1440 °C, respectively. In the process of induction brazing, the heated pipe of AISI 304 steel must not be melted.

As the copper induction coil is water cooled, the material properties of copper for electro-magnetic analysis were entered only for the room temperature with the values: electrical resistivity $\rho_{el} = 1.68\times10^{-8}$ Ω.m and relative permeability $\mu_r = 1$. Water and dry air are similarly as copper non-magnetic, so their relative permeability is also equal to one.

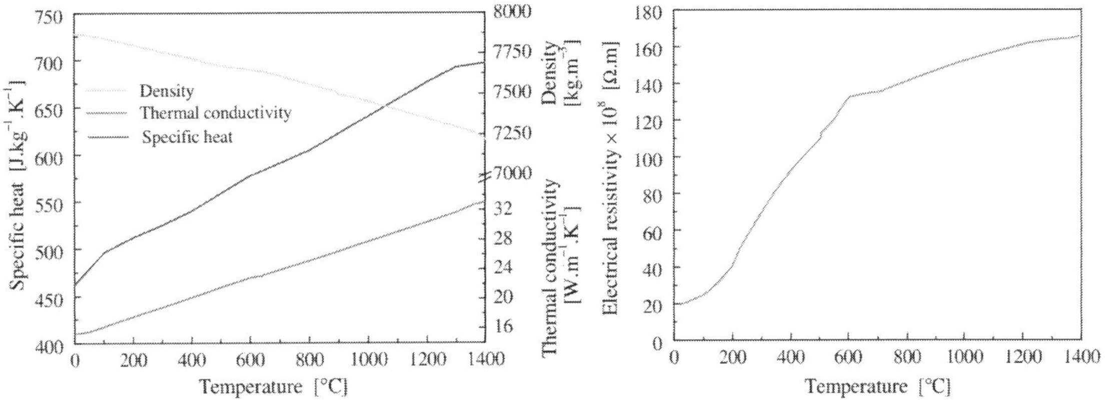

Figure 4. Thermal and electrical properties of the AISI 304 steel in the dependence on temperature.

For electro-magnetic axisymmetric harmonic analysis, the model was loaded by current density in induction coil and current frequency as summarized in Table 2. Moreover, the pipe initial temperature of 20 °C was up-dated based on the results of thermal analysis.

In thermal analysis, the Joule heat distribution computed using electro-magnetic analysis was considered as a loading in the form of the volumetric heat source density. The initial temperature of the heated pipe was supposed to be 20 °C. The pipe cooling by mechanisms of convection and radiation was taken into account and entered as the boundary condition of the 3rd type (Eq. 20). The surrounding temperature $T_f = 20$ °C was defined.

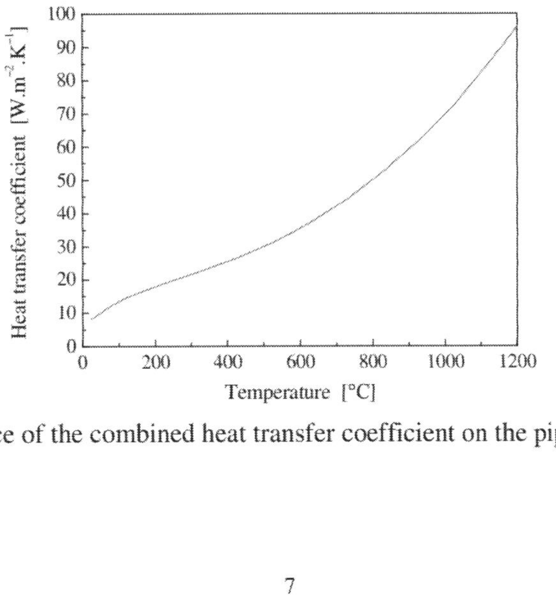

Figure 5. Dependence of the combined heat transfer coefficient on the pipe surface temperature.

The heat transfer coefficient in the dependence on the pipe surface temperature was calculated according to Eq. 21 using classical criterial equations to determine the convection heat transfer coefficient. The obtained dependence of the combine heat transfer coefficient on the temperature is illustrated in Figure 5. In the y-axis, the symmetry conditions were applied for both, electro-magnetic and thermal, analyses by definition of zero A_z magnetic vector potential and zero heat flux.

4.3. Comparison of computed and experimental results

To verify the suggested methodology for numerical simulation of induction heating and developed simulation model, the temperatures measured during experiments were compared with the temperatures computed using numerical simulation in the nodes which location approximately corresponded to the thermocouple positions. In Figures 6 and 7, the time dependence of measured and computed temperatures are plotted exhibiting a good match of the obtained results.

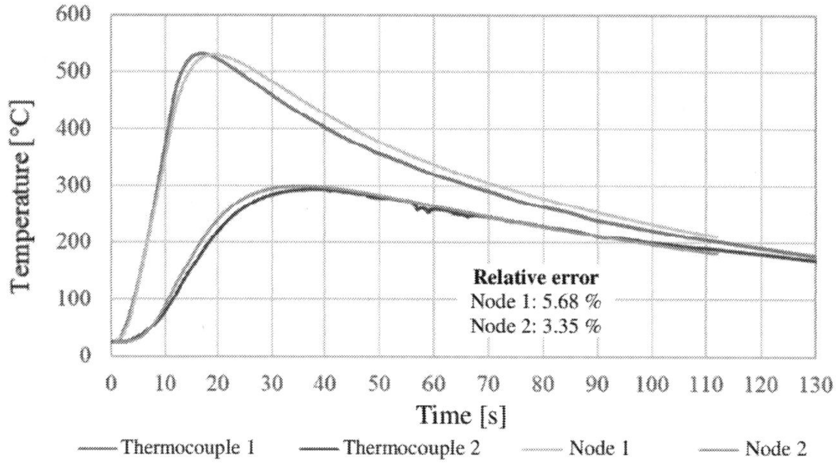

Figure 6. Comparison of measured and computed temperatures during the experiment 1.

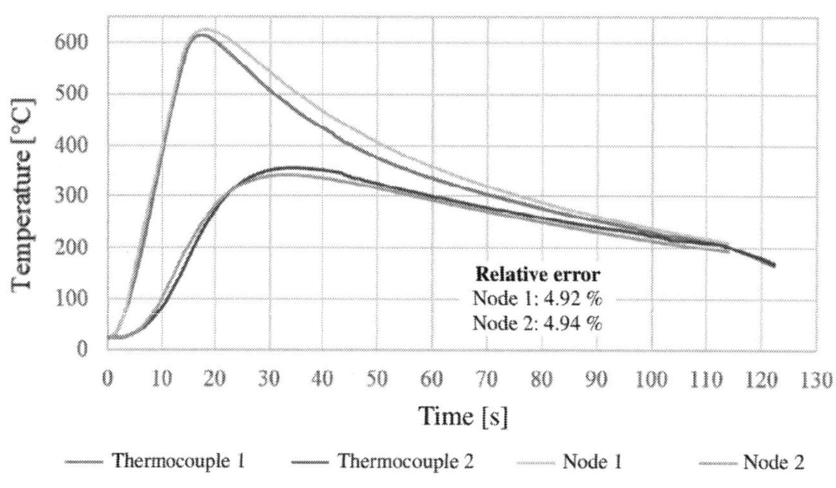

Figure 7. Comparison of measured and computed temperatures during the experiment 2.

Slight deviations of measured and computed temperatures result from the fact that according to generated mesh the positions of nodes do not exactly fit thermocouple locations. Moreover, some deflections could likely arise due to inaccuracies in the material properties and cooling conditions applied. The relative error of computed temperatures corresponding to that measured by thermocouple 1 and thermocouple 2 in the experiment 1 is 5.68 % and 3.35 %, respectively. In the experiment 2, these errors are less than 5 %.

Based on the results of verification experiment and comparison of measured and computed temperatures it can be concluded that the methodology of sequential field coupling applied for numerical simulation of a pipe induction heating using the ANSYS software is correct and can thus be exploited to design parameters of induction brazing of thin-walled pipes.

5. Simulation model for numerical analysis of induction brazing of thin-walled pipes

In the next step, the simulation model was developed in order to analyse the process of induction brazing of pipes made of AISI 304 steel using the copper-based solder.

Geometrical model (Figure 8a) consists of two pipes with the outer diameters of 31.8 mm and 33.6 mm, respectively, the wall thickness of 0.9 mm and the length of 50 mm. The suggested length of the pipe overlap in the brazed joint is 2 mm. The solder has the form of a ring made of the wire 1 mm in diameter. For induction brazing, the same type of induction coil was suggested as applied in verification experiment. The problem was solved as axisymmetric. The generated finite element mesh (Figure 8b) takes into account the skin effect and the concentration of magnetic field in the area of induction coil as illustrated by a detail in Figure 8c.

Material model of the austenitic stainless AISI 304 steel was described in the previous section. The temperature dependence of thermal and electrical properties of the B-Cu97Ni(B) solder computed using JMatPro software is depicted in Figure 9. The melting point of the applied B-Cu97Ni(B) solder is lower. The solidus and liquidus temperatures of the solder were computed to be T_S = 1080 °C and T_L = 1105 °C, respectively. The solder working temperature is ranging from 1115 °C to 1125 °C. The relative permeability of the solder was supposed to be 1.35.

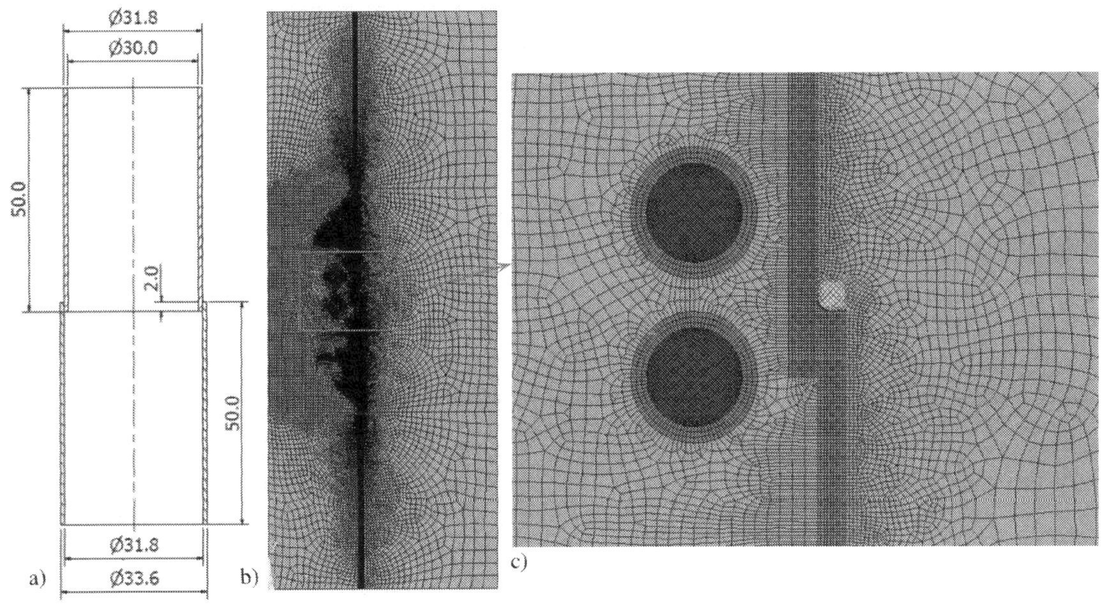

Figure 8. Geometrical model (a) and finite element model (b) for induction brazing of pipes with a detail of generated mesh near the induction coil (c).

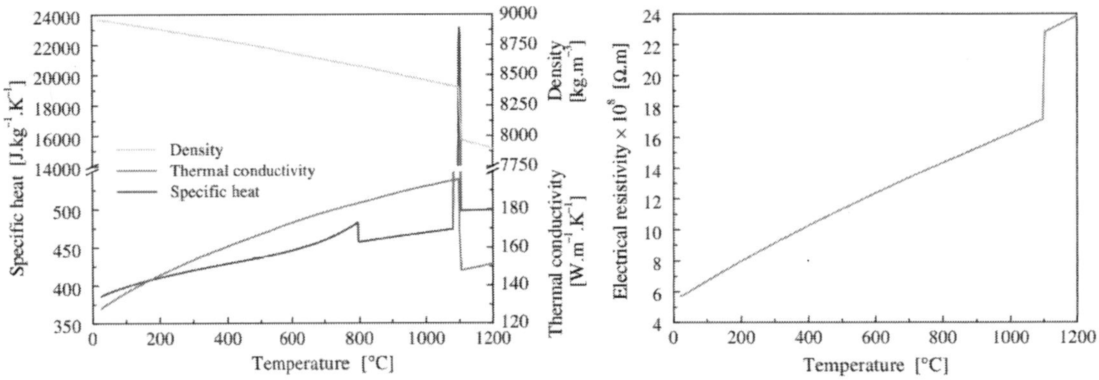

Figure 9. Thermal and electrical properties of the B-Cu97Ni(B) solder in the dependence on temperature.

Similarly to the verification model, symmetry conditions were defined in the *y*-axis applying zero magnetic vector potential and zero heat flux density. The pipe cooling by convection and radiation to the surrounding air at the temperature of 20 °C was considered. It was supposed that the thermal and electrical contact between the pipes as well as between the pipe and solder were ideal. The initial temperature of pipes and the solder was 20 °C. The frequency of source current was constant (20 kHz). The current density was varied from the interval of 1.5×10^8 A.m^{-2} to 2.1×10^8 A.m^{-2}.

6. Results and discussion

Using the developed simulation model of the induction brazing, numerical experiments were carried out applying the ANSYS 18.2 software to analyse the electro-magnetic and temperature fields in the thin-walled tubes during the joining process and to design suitable induction heating parameters for the production of a high quality brazed joints.

6.1. Analysis of electro-magnetic and temperature fields

To illustrate the mutual relationships of electro-magnetic variables, the distribution of current density, magnetic vector potential, the intensity of magnetic field, magnetic induction and generated Joule heat at the end of induction heating in the time of 12 seconds applying the current density of 1.7×10^8 A.m^{-2} is shown in Figure 10. In Figure 10f, the temperature distribution at the end of induction heating is documented as well. The intensity of magnetic field is maximal near the induction coil. The highest induced eddy currents flowing through the solder result in the maximum generated Joule heat and the highest temperature increase in the solder. This results can be supported also by Figure 11 illustrating the details of temperature fields in the pipes and solder in the time of 11.96 s when the solder reached the working temperature. The maximum temperature differences in solder at this time are less than 6 °C.

6.2. Influence of current density on the solder temperatures

According to the required heating time of 12 seconds, the influence of current density in induction coil on the temperature fields in the copper-base solder was analysed. In Figure 12, the time histories of the temperatures in the centre of the solder cross-sectional area are depicted for chosen source current densities. The maximal solder temperature is reached at the end of the heating process in the time of 12 s. Consequently, the temperature of the solder decreases owing to the convection and radiation cooling.

In Figure 13, the solder temperature in the time of 12 s as a function of the current density in induction coil is plotted. The temperatures from the interval from 1115 °C to 1125 °C correspond to the recommended working temperatures of the B-Cu97Ni(B) solder.

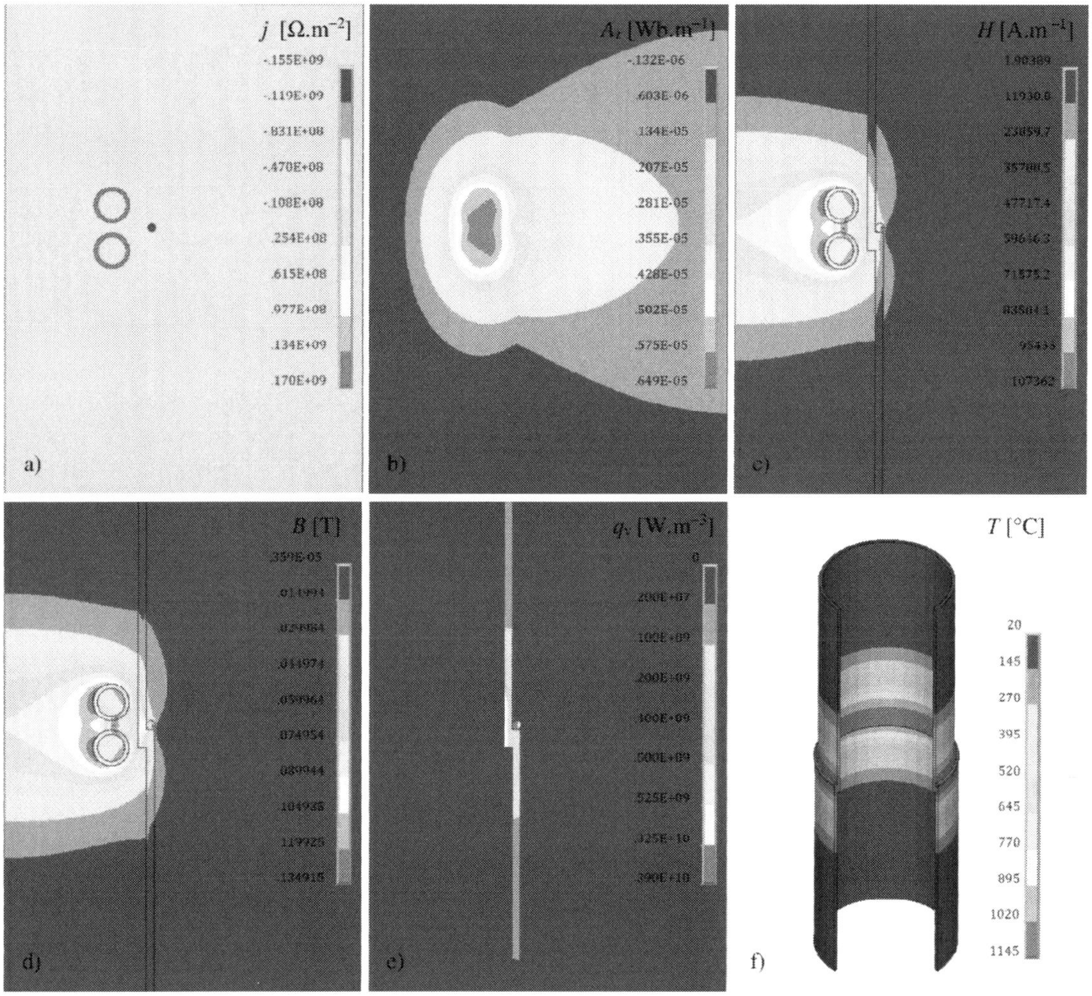

Figure 10. Details from the distribution of electro-magnetic variables in the time of 12 seconds a) current density, b) vector potential, c) intensity of magnetic field, d) magnetic induction, e) generated Joule heat and f) temperature field (source current density of 1.7×10^{8} $A.m^{-2}$).

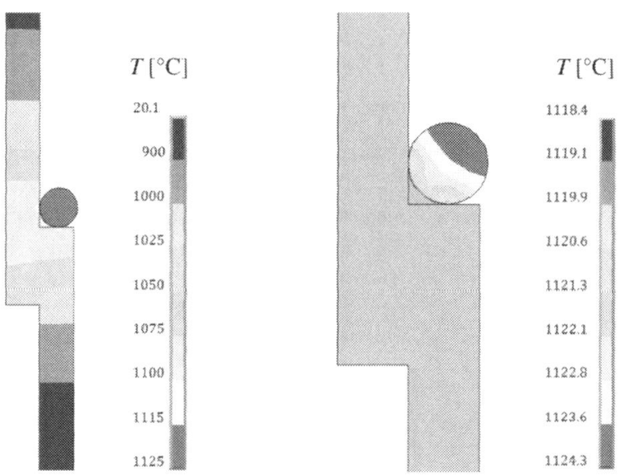

Figure 11. Details of temperature fields in the pipes and solder in the time of 11.96 seconds.

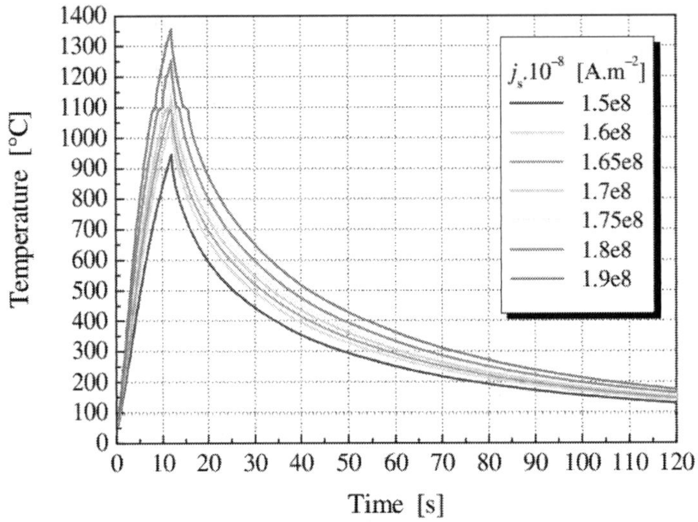

Figure 12. Time histories of the temperatures in the centre of the solder cross-sectional area for chosen source current densities.

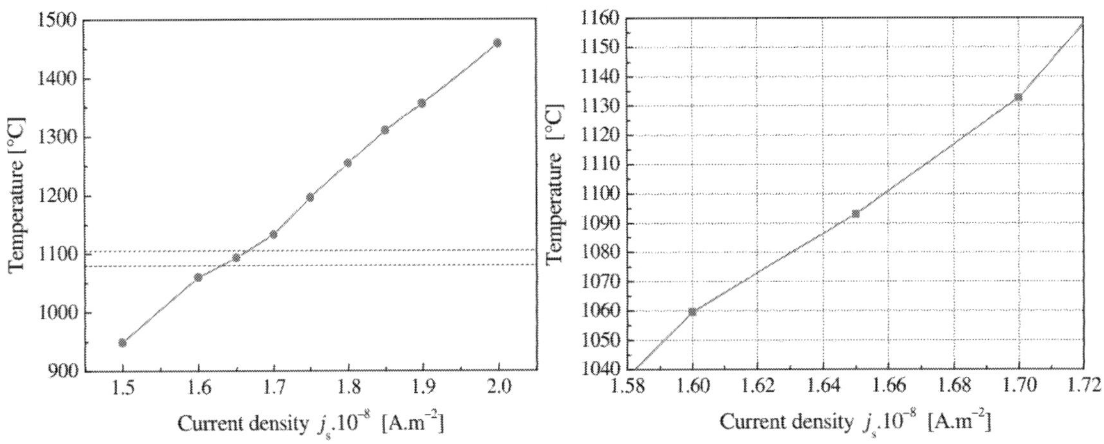

Figure 13. Solder temperature in the time of 12 sesonds in the dependence on the current density in induction coil.

Based on the results presented in a detail from Figure 13, the current density range from 1.67×10^8 A.m^{-2} to 1.69×10^8 A.m^{-2} can be proposed for induction brazing of thin-walled pipes.

6.3. Influence of the current density on the heating time of solder to the brazing temperature

In order to design the operating diagram, heating time vs. current density, for induction brazing of thin-walled pipes of the AISI 304 steel using copper-based solder, the effect of the current density on the duration of induction heating of the B-Cu97Ni(B) solder to its liquidus temperature and solder working temperature was investigated.

Figure 14 shows the time dependence of the solder temperature for considered values of the current density in inductor coil with a detail of the time history of temperatures in the region of recommended solder working temperatures. Based on the obtained graphs, the heating times of the B-Cu97Ni(B) solder to the temperature of 1080 °C and 1125 °C, corresponding to the liquidus temperature and maximal recommended solder working temperature, respectively, were determined and plotted to the operating diagram in Figure 15. Using the operation diagram, the heating time needed to attain required solder temperature applying selected current density can be defined or vice versa for a desired heating time, the required current density can be determined.

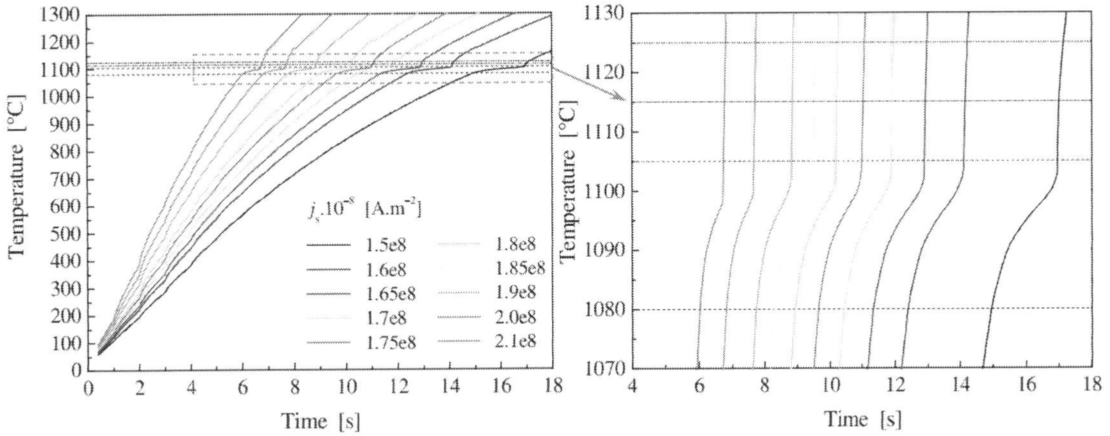

Figure 14. Time dependence of the solder temperature for chosen values of the current density in inductor coil with a detail of the time history of temperatures in the region of recommended solder working temperatures.

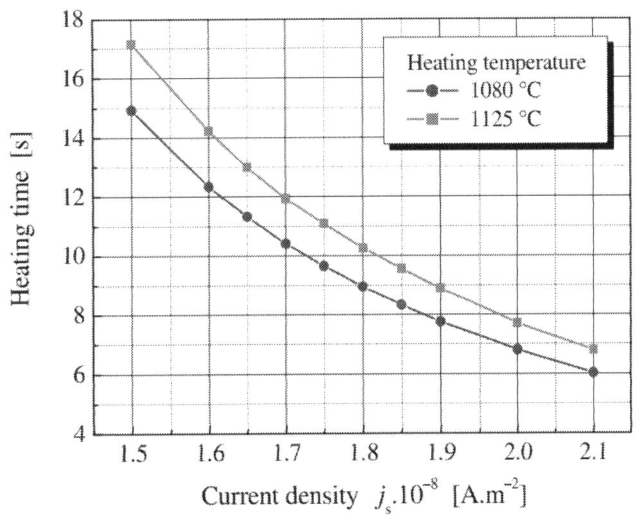

Figure 15. Operating diagram for induction brazing of stainless steel pipes.

7. Conclusions

The research was focused on the investigation of the process of induction brazing of thin-walled pipes made of AISI 304 steel using the B-Cu97Ni(B) solder with the aim to design appropriate parameters of induction brazing by means of numerical simulation of the process in the ANSYS software.

- For induction heating of pipes, a two-threaded copper water cooled induction coil was suggested.
- Simulation model for numerical analysis of pipe induction brazing was developed.

- The process of induction brazing was considered as a coupled problem involved harmonic electro-magnetic analysis and transient thermal analysis.
- To verify the developed simulation model of induction brazing, verification experiment was performed.
- Subsequently, the simulation model was applied to design suitable parameters for the induction brazing of thin-walled AISI 304 steel pipes using a copper-based solder.
- Based on the obtained results, the operating diagram (current density vs. heating time) for the investigated induction brazing process was developed.

Acknowledgement
The research has been supported by the Scientific Grant Agency of the Slovak Republic within the project VEGA No. 1/1010/16, KEGA No. 029STU-4/2018 and ITMS 26220120048 project under the Research & Development Operational Programme funded by the ERDF.

References
[1] Roberts P M 2016 *Introduction to Brazing Technology* CRC Press London NY ISBN 9781498758444
[2] Ainali M Ed. 2008 *Brazing Handbook* CuProBase Västerås Sweden
[3] The Brazing Guide. Available at: http://www.ghinduction.com/wp-content/uploads/2011/09/GH-Brazing-Guide1.pdf (Accessed 10.08.2018)
[4] An Introduction to Brazing. Fundamentals, Materials, Processing. Available at: https://www.oerlikon.com/ecomaXL/files/metco/oerlikon_BRO-0010.4_Introduction_to_Brazing_EN.pdf (Accessed 12.06.2018)
[5] Rudnev V 2005 *Joining components by induction heating* Available at: https://inductoheat.eu/wp-content/uploads/sites/15/2015/04/Joining_by_induction__part_1_Optimieren.pdf (Accessed 12.06.2018)
[6] Rudnev V, Loveless D and Cook R L 2017 *Handbook of Induction Heating* CRC Press ISBN 9781138748743
[7] Pimenta G and Jarman R A 2010 Corrosion of Metal Joints In *Reference Module in Materials Science and Materials Engineering, Shreir's Corrosion* Elsevier **3** 2447-2462 https://doi.org/10.1016/B978-044452787-5.00121-9
[8] Davis J R Ed. 2006 Corrosion of Austenitic Stainless Steel, In *Corrosion of Weldments* ASM International p. 43 ISBN: 978-0-87170-841-0
[9] Chen Y and Cui H 2018 *Materials* **11** 1586 https://doi.org/10.3390/ma11091586
[10] Xiaowei Wu et al 2000 *Mater Sci Eng* A **288** 1 84-90
[11] Łęcka K M et al 2016 *J Laser Appl* **28**, 032009 https://doi.org/10.2351/1.4948726
[12] Szubzda B *et al* 2016 *IOP Conf. Ser.: Mater. Sci. Eng.* **113** 012017
[13] Zaletelj H, Fajdiga G and Nagode M 2013 *Mater Sci Eng* A **564** 478–484
[14] Wei Z et al 2016 *Failure mechanisms and modes analysis of vehicle exhaust components and systems* In Handbook of Materials Failure Analysis with Case Studies from the Aerospace and Automotive Industries, Butterworth-Heinemann, 393-432 ISBN 9780128009505
[15] Evin E, Tomáš M and Výrostek M 2016 *Acta Mechanica et Automatica* **10** 3 189–194.
[16] Doche M L et al 2006 A novel accelerated corrosion test for exhaust systems by means of power ultrasound *Corros Sci* **48** 12 4080-4093
[17] Sugiyama M, Watanabe I and Murofushi K 2016 *Stainless Steel Automotive Fuel Pipe.* United States: Patent Application Publication. 24. November 2016. Patent: US 2016/0341163 A1

[18] Ellbruck S 2017 *Handbook on consumables for welding, brazing and spraying* Fontargen GmbH, Eisenberg. Available at: http://www.bohler-uddeholm.cz/media/ Prirucka_Fontargen _AJ.pdf (Accessed 12.06.2018)
[19] Fontargen Brazing Product Catalog. Available at: https://www.voestalpine.com/welding/ Brands/Fontargen-Brazing (Accessed 12.06.2018)
[20] AISI Type 304 Stainless Steel. Available at: http://asm.matweb.com/search/Specific Material.asp?bassnum=mq304a (Accessed 19.05.2018)
[21] Behulova M, Mašek B and Meyer L W 2006 *Materialprufung* **48** 5217-224
[22] Incropera P F and De Witt D P 1996 *Fundamentals of heat and mass transfer* (New York: J. Wiley and Sons)
[23] ANSYS® Academic Research Mechanical, Release 18.1
[24] JMatPro Help, Release 6.1 2012 Sente Software Ltd.

IOP Conf. Series: Materials Science and Engineering **465** (2019) 012002 doi:10.1088/1757-899X/465/1/012002

Design of the Time of Flight ERDA system for 6 MV tandem ion accelerator

J Dobrovodský, D Vaňa, M Strémy and M Beňo

Slovak University of Technology in Bratislava, Faculty of Materials Science and Technology in Trnava, Advanced Technologies Research Institute, Jána Bottu 25, 91724 Trnava, Slovakia

jozef.dobrovodsky@stuba.sk

Abstract. At present, the Ion Beam Laboratory MTF STU in Trnava [1] uses the basic Ion Beam Analysis (IBA) methods, namely Rutherford Backscattering Spectrometry (RBS), Particle Induced X-ray Emission (PIXE), in limited extend Nuclear Reaction Analysis (NRA) and Elastic Recoil Detection Analysis with He beam (He ERDA). The development of the Heavy Ion Elastic Recoil Detection Analysis (HI ERDA) system for our laboratory was initiated. The supposed analysing projectiles will be from O to Au with energies ~ 1 MeV/u. HI ERDA is ideal for measuring the depth concentration profiles of all elements lighter than the mass of the primary ions, present in near surface layers in heavy base material. During the HI ERDA experiment, the high energy heavy ions impinge on the sample at an acute angle, atoms of the sample are elastically recoiled, and their energy and mass has to be determined. We have chosen the detection system consisting of the Time of Flight (ToF) telescope in combination with silicon charge particle detectors. Two time gates will be used to acquire the two time stamps. The fast acquisition electronics has to collect the coincidence energy/ToF/yield multi parametric spectra. The proposed ToF ERDA signal processing system consists from combination of analogue nuclear electronics and the fast digital multi-parameter acquisition system. The first draft of the ToF HI ERDA for Trnava IBA laboratory is presented.

1. Introduction

The Ion Beam Laboratory of the Slovak University of Technology (STU) with the 6 MV Tandem Accelerator currently operates experimental end station dedicated for Ion Beam Analysis [1], which is equipped for the basic IBA methods, specifically they are RBS, PIXE, to a certain extend also resonant and non-resonant NRA [2] and He ERDA.

RBS is ideal for determination of heavy elements depth profiles in light substrate, alternatively resonant and non-resonant NRA can measure depth distribution of individual elemental isotopes of light elements ($Z < 15$) and PIXE is limited to elements heavier then Al.

Determination of light elements like oxygen, carbon or nitrogen in or on heavy substrates which can be found in many compounds is challenging with those listed IBA techniques. RBS has relatively low sensitivity to light elements in heavier matrices. He ERDA enables to determine the depth distribution of hydrogen and its isotopes. NRA can usually determine only one element at the time and the analysis of the other elements has to be obtained by separate subsequent measurements. It can reach an excellent depth resolution and good sensitivity, but the measurements are time consuming even for single radionuclide.

With the increasing development of multi-layer thin microelectronic devices, thin magnetic storage devices, metallized polymers, advanced ceramic materials, and tribo-coatings, etc., there is a rapidly increasing need to measure depth profiles of elements in surface layers from 10 to 1000 nm thickness.

Effective determination of elemental depth profiles of all elements, including light elements can be performed by Heavy Ion Elastic Recoil Detection Analysis. ERDA is a complementary analytical method to RBS. HI ERDA is an effective method for quantitative depth profiling of all elements in the sample surface region simultaneously within a single measurement in just a few minutes [3].

2. Elastic Recoil Detection Analysis

Conventional He ERDA with a primary helium ion beam and HI ERDA - high-energy heavy ion are considered the two main versions of ERDA. A typical detection configuration for He ERDA is the solid state detector (SSD) - e. g. silicon charge particle detector with an absorber foil placed in front of it. The lighter recoiled protons penetrate through the foil to the detector, while the scattered primary He ions are absorbed in the foil.

2.1. HI ERDA

In HI ERDA heavy projectiles at energies typically of tens MeV collide with sample atoms and knock the atoms lighter than the primary ions out of the sample in the forward direction. A velocity as well as energy are measured simultaneously, and mass of recoil particle is identified from the relation between energy and velocity. Since no stopper foil is used, depth resolution is better than in the case of conventional ERDA.

HI ERDA was successfully applied for the first time more than 40 years ago at the University of Montreal Tandem Accelerator for determination of the depth distribution of light elements in heavy materials [4]. In this case the ^{35}Cl incident ions in energy range from 30 to 40 MeV were chosen to reach an optimum depth resolution.

The acquisition and evaluation of energy spectra of the recoiled atoms is more complex problem. Preferably, it is appropriate to use the primary beam at the energies around 1 MeV/u, where the kinematics of the atom collisions and the scattering cross sections are well known and the stopping powers are near the maximum (which contributes to the high-depth resolution). Typical ions used for HI ERDA are from ^{16}O through ^{35}Cl, ^{63}Cu and ^{79}Br, ^{127}I to ^{197}Au, accelerated at energy from tens, up to few hundreds of MeV. Energies required for these beams are out of reach of many IBA laboratories. [3].

2.2. ERDA detection systems

In HI ERDA experiment the recoiled sample atoms and scattered incident ions can be separated either by their nuclear charge or by their mass. To achieve this separation, various types of ΔE-E detectors, gas ionization chamber (GIC) and TOF telescopes can be used.

In ΔE – E detection system, the energy deposition (or the stopping powers) and the total energy of atoms are detected. It uses the fact that the stopping power of ions scales as Z, so the different elements will slow down at different rates, generating charge carriers. The gas filled chamber can be used as ΔE – E detector, alternatively E can be measured by SSD detector.

Time-of-Flight Elastic Recoil Detection Analysis (ToF ERDA) telescope usually consists of two timing detectors combined with an energy detector, see figure 1. ToF ERDA was developed in the early 1980s for light element profiling [5, 6].

A velocity as well as energy are measured simultaneously. The recoil masses are identified by means of a coincident measurement of the particle velocity and the total energy [7]. A recoils originating from deeper in the sample will have less energy, or correspondingly longer time-of-flight, than recoils from the surface. After evaluation of recorded multi-parameter spectra, the depth profiles of all individual elements present in the sample can be determined.

The timing detectors measure the time-of-flight over a fixed distance. Although only the ToF spectra are used for further depth profile analysis, mostly energy measurement restrict the mass resolution of the whole system.

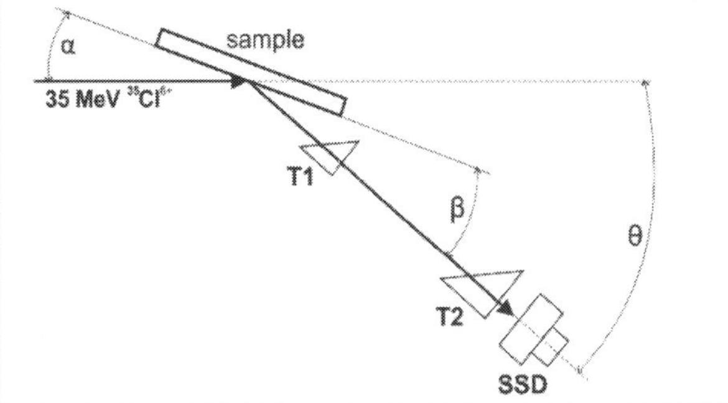

Figure 1. The principle ToF ERDA, the primary beam of 35 MeV $^{35}Cl^{6+}$ is incident on the sample at an angle α, the forward scattering angle is θ. The detection part, the ToF ERDA telescope consists of T1 and T2 timing gates (detectors) and the SSD energy detector.

SSD or GIC detectors are typical alternatives for ions total energy measurement. In both cases their intrinsic resolution limits mass separation below a few MeV recoil energy. Moreover, SSD detectors suffer considerable radiation damage. The energy resolution of the Si charged particle detectors for heavy ions is significantly worse than for light ions like H and He. As SSD detectors start to degrade after about 10^8 fission fragments/cm^2, their lifetime can be considerable limited [8]. The real lifetime of detectors depends also on the experimental conditions and the composition of the analyzed samples. The advantages of SSD detectors are simplicity of handling and operation.

3. ToF ERDA system for STU laboratory

One of the latest installation of the new ToF ERDA telescope was realized at the University of Montreal, where the original telescope was upgraded [9]. In their case the time-of-flight unit consists of two detectors that use a thin carbon membrane and microchannel plates (MCP) to produce the start and stop signals. The time is used to measure the energy with excellent resolution. The timing units were designed by the Jyväskylä group [10]. The main components of the timing units are the 10 mg/cm2 carbon foils; which emits and accelerates the electrons when an ion passes through it; the ''toblerone", which forms electric field-free volume for the electrons; the electrostatic mirror, which reflects the electrons towards the base of the unit; the microchannel plates (MCP), which multiplies the electrons and the anode, which collects the electron signal and produces a fast, ns pulses.

Further processing of these pulses allows to achieve a resolution bellow 200 ps in ToF measurement. The time between the first and second time signal ranges from tens to hundreds of ns. The time measurement depends only on the separation between the timing C foils, delays in the timing detectors and connecting cables, and will not depend on the detected element and detector damage.

On the other hand, semiconductor detectors have an energy ion-depend calibration that depends on the ion, and their resolution worsens with increasing atomic number and they have a problem with pulse height defect (PHD).

3.1. *ToF ERDA telescope*

Our proposal for STU ToF telescope follows the upgraded Montreal design described above with the SSD detector for the energy of the recoils measurement. [9]. The timing detector design will be based on the design by Busch et al. [11], since it offers excellent timing resolution and simplicity of construction.

The HI ToF ERDA system includes the new beamline from the switching magnet to the experimental chamber, equipped with the beam monitoring elements and the HI ERDA detection system based on the ToF telescope, see figure 2.

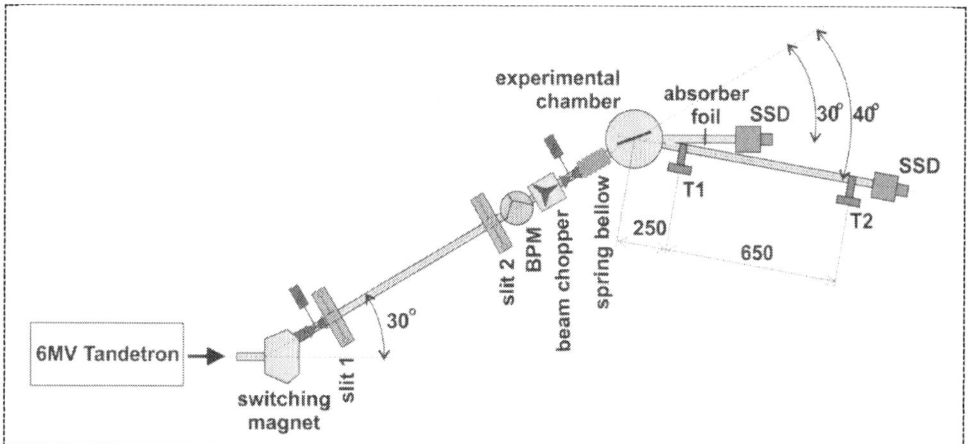

Figure 2. Scheme of the proposed HI ToF ERDA system for STU 6 MV Tandem accelerator, including beam transport from the switching magnet and the ERDA telescope with basic dimensions.

In addition to the ToF telescope another SSD (H) detector, will be used in combination with a stopping foil to monitor the hydrogen, as the detection efficiency of the ToF camera is low for the hydrogen.

3.2. Signal processing

The current collection of data coming from the IBA detectors through NIM ADCs ensures MPA-3 FAST ComTec multi-parameter system. The proposed signal processing of the detection system will combine an analogue front-end electronics with digital multi parameter acquisition system. It will be based on MPA4T FAST ComTec module [12] capable of recording coincident ToF- and energy and also enabling time stamped data acquisition. MPA4T is a combination of a high resolution multi-stop ToF with a multi-parameter data acquisition system. The stop events can be evaluated at a rate of 10 GHz state changes/sec. The system allows to handle up to four NIM ADCs. Multi-parameter spectra can be accumulated in the RAM of the connected PC.

The timing pulses from start/stop detectors will be directed trough a timing preamplifier and a timing discriminator to the fast digitizing time inputs, see figure 3. Signal from SSD (E) energy detector and from the SSD (H) detector will after processing by the standard analog nuclear electronics enter ADCs inputs of the multi-parameter MPA4T acquisition module.

FTP4 can directly recognize coincident data from several detector channels and even very complicated trigger conditions can easily be set by the software during data acquisition or post-processing. [13]. Timestamping offers more options for data analysis as coincidence events can be built also in offline analysis. It can be used for fine-tune the coincidence search parameters to minimize the unwanted background.

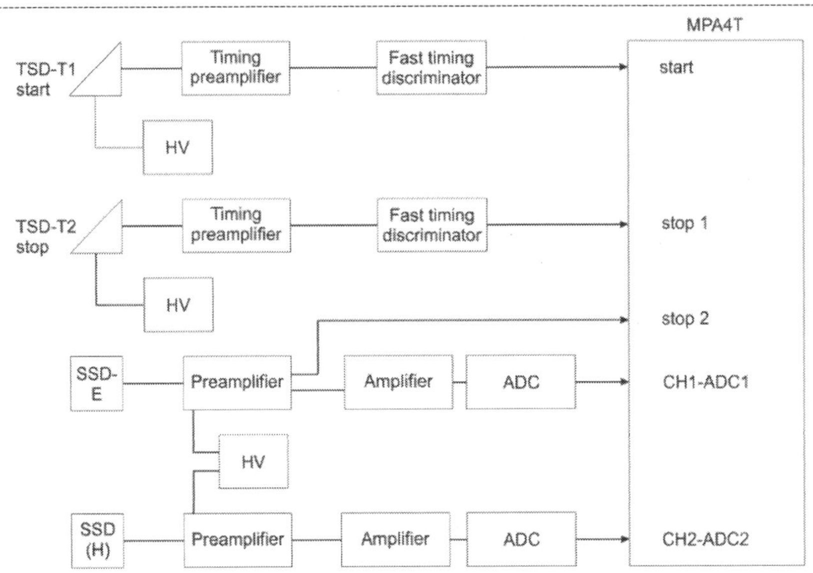

Figure 3. A schematic layout of the fast analogue and digital nuclear electronics for proposed ToF ERDA system.

4. Summary

It is proposed to supplement the existing IBA analytical methods of 6 MV Tandem Accelerator Laboratory by the HI ERDA system. HI ERDA assumes only Coulomb interactions of point charges, the recoiled atoms from the target are directly detected, and there is no significant theoretical difference from RBS. The kinematics in both methods are very similar, as well as the correspondence between the depth and energy loss.

RBS and ERDA are considered to be quantitative depth profiling techniques without standards. The combination of ERDA with standard RBS is a particularly valuable tool. The recoil energy depends strongly on the projectile mass and it is advantageous to use as a heavy ion beam as possible. However, the beam velocity has to remain roughly constant within 0.5-2.0 MeV/u, in order to remain within the Rutherford scattering regime and the main limitation factor is the maximum ion beam energy achievable from an available ion accelerator and the complexity of spectrometers needed for recoil identification [7].

The proposed HI ERDA takes into account the experience from the recently implemented ToF ERDA telescopes. The MPA4T digital multi-parameter acquisition system is considered as one of the most appropriate commercially available system for ToF HI ERDA. It enables ToF with start and 4-channel multi stop signal digitizers with 100 ps timing resolution and a time-stamping multi parameter system with connections of up to 4 or 8 NIM ADC's [12]. Currently the detailed design of individual sub-assemblies of the ToF ERDA analytical system is solved.

Acknowledgments

This work was supported by the Slovak Grant Agency VEGA under contract No. 1/0418/18 and No. 1/0219/16.

References

[1] Noga P, Dobrovodský J, Vaňa D, Beňo M, Závacká A, Muška M, Halgaš R, Minárik S and Riedlmajer R 2017 *Nucl. Instr. Meth. Phys. Res. B*, **409** 264
[2] Ziegler J F, et al. 1978 *Nucl. Instrum. Methods* **14**, 19

[3] Assmann W 1992 *Nucl. Instr. Meth. Phys. Res. B* **64** 267

[4] 'Ecuyer J L, Brassard C, Cardinal C, Chabbal J, Deschênes L, Labrie J P, Terreault B, Martel J G and Jacques R St 1976 *J. Appl. Phys.* **47** 381

[5] Groleau D R, Gujrathi S and Martin J 1983 *Nucl. Instrum. Methods* **218** 11

[6] Thomas J, Fallavier M, Ramdane D, Chevarier N and Chevarier A 1983 *Nucl. Instrum. Methods* **218** 125

[7] Kottler C, Döbeli M, Glaus F and Suter M 2006 *Nucl. Instr. Meth. Phys. Res. B* **248** 155

[8] Siketic Z and Bogdanovic Radovic I 2013 *Nucl. Instr. Meth. Phys. Res. B* **296** 78

[9] Chicoine M, Schiettekatte F, Laitinen M I and Sajavaara T 2017 *Nucl. Instr. Meth. Phys. Res. B* **406** 112

[10] Laitinen M, Rossi M, Julin J and Sajavaara T 2014 *Nucl. Instr. Meth. Phys. Res. B* **336** 55

[11] Busch F, Pfeffer W, Kohlmeyer B, Schüll D and Pühlhoffer F 1980 *Nucl. Instr. Meth.* **171** 71

[12] MPA4T: Multiparameter TOF System, FastComTech website, URL: *https://www.fastcomtec.com/fwww/datashee/mpa/mpa4t.pdf*

[13] Rossi M, Rahkila P, Kettunen H and Laitinen M 2015 *Nucl. Instr. Meth. Phys. Res. B* **347** 39

Preparation of Er-doped $Yb_3(Al, Fe)_5O_{12}$ garnets

V Jakeš[1], K Rubešová[1], J Havlíček[1], V Polák[1], D Sedmidubský[1] and J Oswald[2]

[1] University of Chemistry and Technology Prague, Technická 5, 166 28 Prague, Czech Republic

[2] Institute of Physics of the Czech Academy of Sciences, Cukrovarnická 10, 162 53 Prague, Czech Republic

Vit.Jakes@vscht.cz

Abstract. This work focuses on mixed YbAG/YbIG garnet in a progressive substitution Al/Fe, where aluminium and iron occupy the tetrahedral or octahedral cation sites. In addition to it, the prepared garnet samples were doped with erbium that can profit from ytterbium-based matrix increasing so its absorption cross section. Bulk samples of Er-doped ytterbium–aluminium garnet ($Yb_{2.94}Er_{0.06}Al_{5-x}Fe_xO_{12}$; x = 0–5) were prepared by a sol-gel method based on Pechini process with citric acid and ethyleneglycol as polyesterification agents. The annealing had to be carried out in oxygen atmosphere in order to stabilize Fe^{III} in the phase equilibrium. Photoluminescence in the range of 1450–1650 nm was measured and the influence of iron doping on erbium luminescence properties was established.

1. Introduction

Rare earth iron garnets, variably doped with another rare earth ion, are frequently investigated for use in photonic integrated circuits. These magneto-optical oxides, such as $Ce:Y_3Fe_5O_{12}$ or $Bi:Y_3Fe_5O_{12}$, possess high Faraday rotation and low optical loss, thereby making them applicable as optical insulators in photonic waveguides for telecommunications [1].

Iron garnets are part of a large family of mixed oxide ceramics with a cubic crystal structure (space group Ia-3d). Their general formula is $A_3^{+III}Fe_5^{+III}O_{12}$, where A = Y, Bi or a rare earth metal [2]. Yttrium-iron garnets ($Y_3Fe_5O_{12}$ or YIG) were discovered in 1957 [3], since which time iron garnets have been widely studied for their magnetic or electrical properties (ferrimagnetism, high electrical resistivity and thermal conductivity, etc.) [4]. The substitution of other metals at the Fe site in the YIG structure led to a change in the magnetic and electrical properties of the material; for instance, the partial substitution of aluminium for iron was found to decrease the Curie temperature [5]. While $Y_3Fe_5O_{12}$ remains the most extensively studied iron garnet, with substitutions being performed at both yttrium and iron sites, papers published on other members of the iron garnet family (e.g. $Yb_3Fe_5O_{12}$, $Ho_3Fe_5O_{12}$ or $Tb_3Fe_5O_{12}$) are rather scarce and focus only on their magnetic properties [6-8].

To extend the use of iron garnets in photonic applications, they can be further doped by an activating luminescent ion. Doping with erbium is suitable for use in the third telecommunication window because of its photoluminescence at around 1530 nm. When the photoluminescence of Er-doped $Y_3Fe_5O_{12}$ and $Bi_3Fe_5O_{12}$ thin films was investigated [2], the Er^{3+} emission bands at 540 and 980 nm disappeared while the bands in the range 1530–1560 nm remained intensive. This was due to both strong absorption in the broad charge transfer band at lower wavelengths and the sensitizing effect between the $^4T_{1g}$ energy level of Fe^{3+} and the $^4I_{11/2}$ energy level of Er^{3+}. Co-doping with ytterbium can further enhance the absorption cross-section of Er^{3+} ions. The energy transfer from the ytterbium $^2F_{7/2}$ energy level to the erbium $^4I_{11/2}$ energy level is well known [9] and was also tested in ytterbium aluminium garnets in which ytterbium was both a co-dopant and a constituent part of the matrix [10, 11].

In terms of magnetic properties, exchange interactions were described for undoped $Yb_3Fe_5O_{12}$ (YbIG) [12-14] and their influence on the splitting of Yb^{3+} energy levels in YbIG was reported [8, 15]. For the Yb_2O_3–FeO–Fe_2O_3 system [16, 17], phase equilibria were described and four stable ternary phases identified: $YbFeO_3$, $YbFe_2O_4$, $Yb_3Fe_5O_{12}$ and $Yb_2Fe_3O_7$.

However, to our best knowledge, no paper has yet been published about the optical and magnetic properties of Er-doped ytterbium iron or ytterbium iron-aluminium garnets. Here, we present series of Er-doped ytterbium aluminium-iron garnets. The effect on properties is studied across the complete concentration range of Al-Fe substitution in bulk samples with a fixed erbium content: $Yb_{2.94}Er_{0.06}(Al, Fe)_5O_{12}$. To ensure the homogeneous distribution of dopants, the samples were prepared using a sol-gel method based on Pechini polyesterification. The samples were compared by measuring the Er^{3+} photoluminescence properties and evaluating their magnetic characteristics.

2. Experimental

Preparation of samples

A set of bulk samples of a general stoichiometry $Yb_{2.94}Er_{0.06}Fe_{5-x}Al_xO_{12}$ (x = 0, 1, 2, 3, 4, 5) was produced from precursors prepared by a sol-gel process. These samples have a constant content of erbium dopant and differ in the ratio of iron and aluminium atoms; they are labelled according to this ratio (e.g. $Yb_{2.94}Er_{0.06}Fe_4Al_1O_{12}$ is F4A1, etc.)

The method chosen for the precursors' preparation was based on the Pechini patent. As a source of cations, $AlCl_3.6H_2O$ (p.a., Honeywell), $FeCl_3.6H_2O$ (p.a., Sigma-Aldrich), $Yb(CH_3COO)_3.xH_2O$ (99.9%, Strem Chemicals) and $Er(CH_3COO)_3.xH_2O$ (99.9%, Strem Chemicals) were used. Citric acid (p.a., Lach-Ner) and ethylene glycol (p.a., Penta) were used for the polyesterification reaction. The chlorides were dissolved in water to a concentration of approx. 10 g of Al per L and 15 g of iron per L; the actual concentration was measured by means of gravimetry. The acetates were dehydrated by heating at 200 °C for 12 hours prior their use.

First, citric acid was dissolved in the aqueous solution of Al^{3+}. The aqueous solution of Fe^{3+} was then added, followed by solid dehydrated acetates of ytterbium and erbium. After all solid was completely dissolved, ethylene glycol was added and the mixture was magnetically stirred for 30 minutes. Water was evaporated at 85 °C and then the temperature of the mixture was elevated to 130 °C to start a polyesterification process. Amorphous gel was then dried for 2 hours at 240 °C and decomposed at 600 °C (2 hours) and 900 °C (1.5 hours). The precursor was pressed into pellets and sintered at 1250 °C in oxygen atmosphere for 144 hours (the effect of annealing atmosphere was also studied as will be discussed in the Results section).

Characterization

The phase composition of the prepared films was determined by XRD analysis. Data were collected at room temperature with Bruker-AXS D2 Phaser powder diffractometer with parafocusing Bragg–Brentano geometry using CoK_α radiation in a range of 10–100° 2θ with a step size of 0.02°. The phase analysis was performed in the HighScore Plus program using the PDF-4 database.

For the photoluminescence (PL) measurement, the excitation radiation of a wavelength of 980 nm emitted by a semiconductor laser POL 4300 was used and the light was processed by a double grated SLD-1 monochromator. Luminescence was detected by a Ge detector J16-2TE (Teledyne Judson Technologies). The spectra were measured in a wavelength region of 1450–1650 nm that corresponds to the Er^{3+} transitions from the manifold $^4I_{13/2}$. All measurements were carried out at room temperature. For evaluation, all luminescence spectra were transposed to the base level and normalized to a standard sample (an Er^{3+}/Yb^{3+}:$LiNbO_3$ single crystal used as a reference).

3. Results

After the decomposition of precursor powder and prior to annealing, an XRD analysis of precursor phase composition was performed; the results for the sample F5A0 ($(Yb,Er)_3Fe_5O_{12}$) are shown in figure 1.

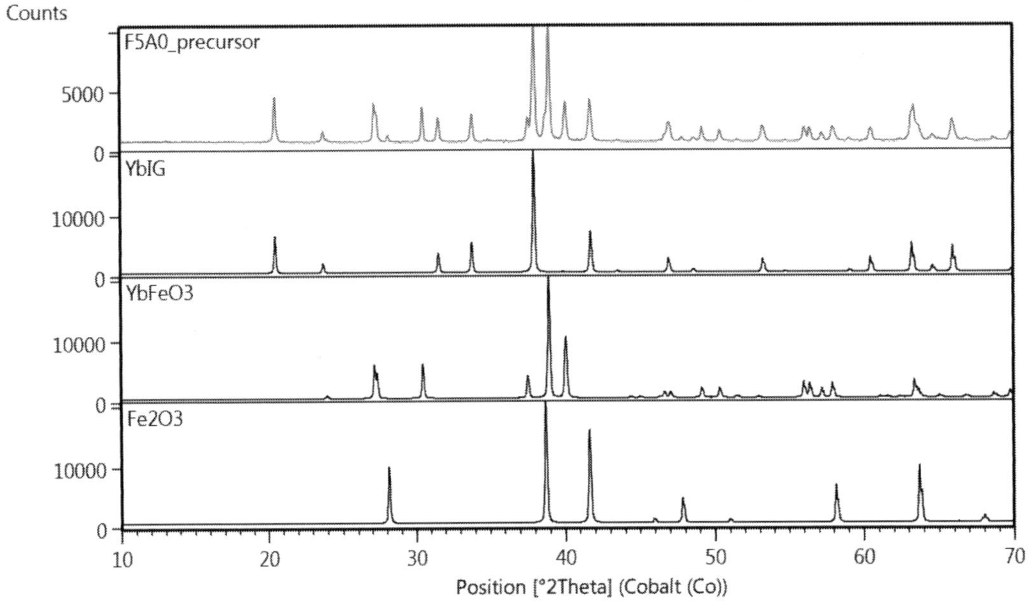

Figure 1. XRD patterns of the precursor powder (F5A0) and reference patterns of the identified phases ($Yb_3Fe_5O_{12}$, $YbFeO_3$ and Fe_2O_3).

As can be seen, the precursor contained the desired garnet phase, together with ytterbium-iron perovskite (YbIP) and iron oxide (hematite) already after the gel was decomposed. Here, the advantage of producing highly reactive particles via sol-gel methods is apparent – garnet is formed after 1.5 hours at as low a temperature as 900 °C, while it would need to be annealed for much longer time and at a temperature almost twice as high if solid state reaction was used.

The effect of the annealing atmosphere on the phase composition of final samples was studied as well. The precursor F5A0 was pressed into pellets that were then annealed in air at 1300 °C for 3 days or in oxygen flow at 1200 °C for 4 days. The phase composition of the prepared samples was determined with XRD analysis again (a section of 20–50° of 2theta is shown in figure 2).

Even though the precursor powder contained the YbIG garnet phase, after it was annealed in air, most of it decomposes to YbIP perovskite, Yb_2O_3 and iron oxides (Fe_2O_3 and Fe_3O_4 – pattern not included in figure 2). This effect corresponds to the fact that Fe_2O_3 tends to decompose at high temperature with a simultaneous reduction of Fe^{3+} to Fe^{2+} [18]. In oxygen flow, on the other hand, almost single phase YbIG was obtained with a small fraction of Fe_2O_3 impurity. In order to maintain iron cations in their higher oxidation state, an increased partial pressure of oxygen is necessary. This is why oxygen atmosphere was introduced when the whole series was annealed.

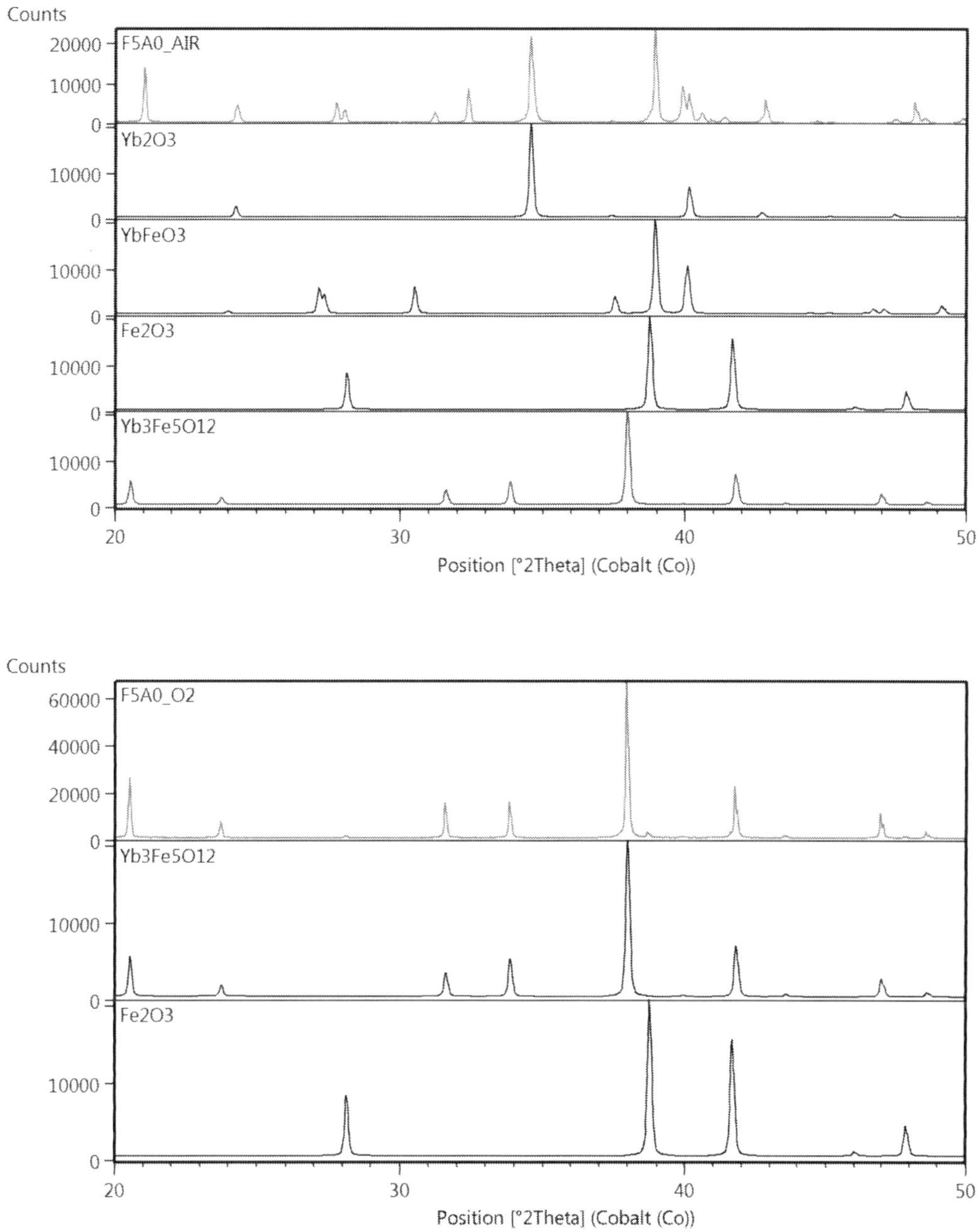

Figure 2. XRD patterns of the sample F5A0 annealed in air (left) or oxygen (right).

Figure 3 shows XRD patterns of all prepared samples. Since Fe^{3+} cation is slightly larger than Al^{3+}, with the increasing content of iron cations the lattice cell parameter increases as well and garnet structure reflections shift towards smaller 2theta value. All samples contained a small fraction of Fe_2O_3 impurity that, based on the Rietveld refinement and subsequent phase analysis, was determined to be around 7.5%. In the Yb/Y-Fe-Al-O system, there are probably competitive processes – garnet vs perovskite thermodynamic stability, the decreasing stability of Fe^{III} with temperature etc. – causing difficulties in the pure-phase YbIG preparation. Nevertheless, for the evaluation of Er^{3+} luminescence properties in the $Yb_3(Fe,Al)_5O_{12}$ matrix, the presence of Fe_2O_3 phase does not play any role.

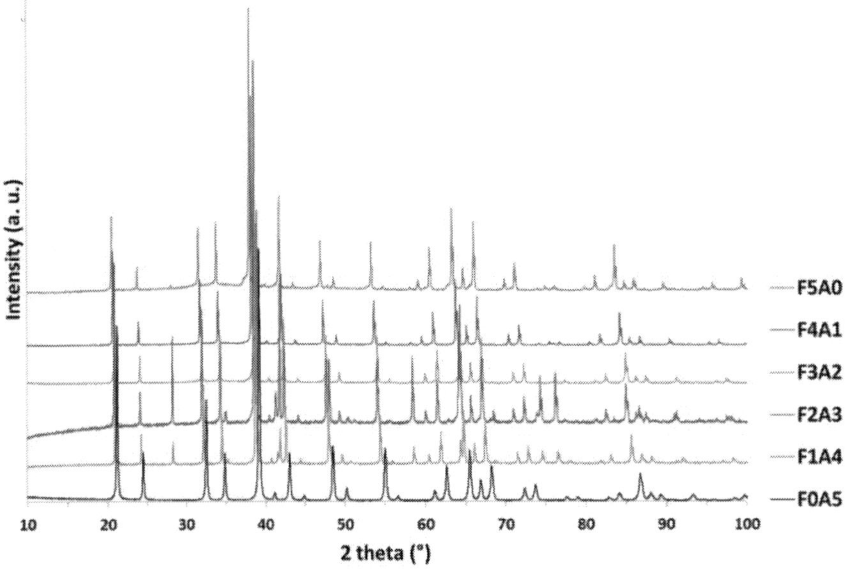

Figure 3. XRD patterns of all prepared samples

The room temperature emission of erbium cation ($^4I_{13/2} \rightarrow {}^4I_{15/2}$ energy transfer) was measured in a range of 1450–1650 nm; the results of this photoluminescence measurement are shown in figure 4. The sample F0A5, that contains no iron, shows discretely resolved fine splitting that corresponds to the transitions from the $^4I_{13/2}$ manifold and demonstrates homogeneous distribution of doped Er^{3+} cations in the garnet matrix and their regular crystal vicinity. When iron is introduced into the garnet matrix, this fine splitting disappears, but, up to 40 % doping of Fe, the PL intensity is comparable with the sample without any iron introduced. The luminescence intensity decreases until it is almost completely lost in samples with the highest content of iron, as shown in the inset of figure 4. The decrease (or disappearance in the case of the highest Fe doping) is probably caused by an energy transfer between Fe^{3+} and Er^{3+} energy levels that suppresses the radiative transition of erbium ions.

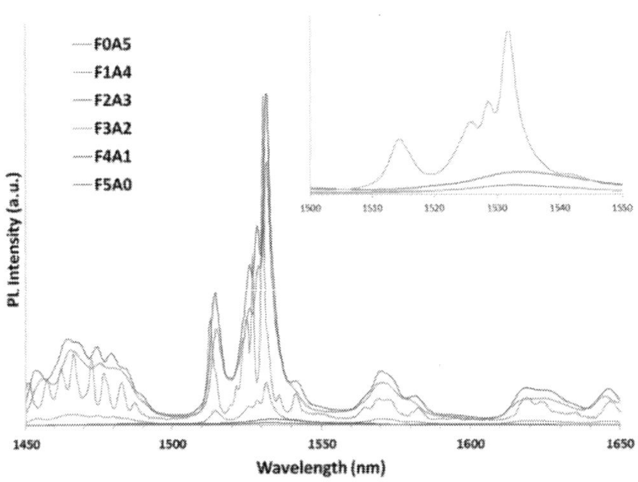

Figure 4. Photoluminescence of the samples
(samples with weaker emission (F3A2–F5A0) shown in the inset)

4. Conclusion

Bulk samples of a general stoichiometry $Yb_{2.94}Er_{0.06}Fe_{5-x}Al_xO_{12}$ (x = 0–5) were prepared via Pechini-based polyesterification. The oxygen atmosphere was used to stabilize the higher oxidation state of Fe and to prevent the decomposition of the garnet phase during annealing. Prepared samples were almost single phase garnets with a small fraction of Fe_2O_3 impurity. Erbium luminescence was studied in the ytterbium garnet matrix. With the increasing content of iron, the intensity of emission is comparable with an un-doped sample up to 40 % iron content and then it decreases down to the full loss of erbium $^4I_{13/2} \rightarrow {}^4I_{15/2}$ energy radiative transition (in the case of 80 and 100 % iron doping).

Acknowledgment

This work was financially supported from specific university research MSMT 21-SVV/2018.

References

[1] Bi L et al 2013 Magneto-Optical Thin Films for On-Chip Monolithic Integration of Non-Reciprocal Photonic Devices *Materials*. **6** 5094
[2] Grishin, A M and Khartsev S I 2012 Luminescent Magneto-Optical Photonic Crystals *Journal of Physics: Conference Series* **352** 012007
[3] Grishin, A M and Khartsev S I 2012 Luminescent Magneto-Optical Photonic Crystals *Journal of Physics: Conference Series* **352** 012007
[4] Azadi Motlagh Z, Mozaffari M and Amighian J 2009 Preparation of nano-sized Al-substituted yttrium iron garnets by the mechanochemical method and investigation of their magnetic properties *Journal of Magnetism and Magnetic Materials* **321** 1980
[5] Grasset F et al 2001 Synthesis, magnetic properties, surface modification and cytotoxicity evaluation of Y3Fe5−xAlxO12 (0⩽x⩽2) garnet submicron particles for biomedical applications *Journal of Magnetism and Magnetic Materials* **234** 409
[6] Opuchovic O, Beganskiene A and Kareiva A 2015 Sol–gel derived Tb3Fe5O12 and Y3Fe5O12 garnets: Synthesis, phase purity, micro-structure and improved design of morphology *Journal of Alloys and Compounds* **647** 189
[7] Sankaranarayanan V K and Gajbhiye N S 1990 Magnetization and magnetic resonance studies of ultrafine Ho3Fe5O12 and Yb3Fe5O12 *Journal of Magnetism and Magnetic Materials* **92** 217
[8] Wickersheim K A and White R L 1960 Optical Observation of Exchange Splitting in Ytterbium Iron Garne. *Physical Review Letters* **4** 123
[9] Auzel F E 1973 Materials and devices using double-pumped phosphors with energy-transfer. *Proceedings of the Ieee* **61** pp 758-786
[10] Hlásek T et al 2016 Sol-gel derived planar waveguides of Er3+:Yb3Al5O12 prepared by a polyvinylpyrrolidone based method *Journal of Sol-Gel Science and Technology* **80** 531
[11] Hlásek T et al 2014 Infrared luminescence in Er3+:Yb3Al5O12 bulk ceramics prepared by sol-gel method *Journal of the European Ceramic Society* **34** 3779
[12] Streever R L 2003 Anisotropic exchange in YbIG *Journal of Magnetism and Magnetic Materials* **263** 219
[13] Tsidaeva N and Abaeva V 2006 The investigation of mechanisms of the magneto-optical activity of rare-earth iron garnet single crystals *Journal of Alloys and Compounds* **418** 145
[14] Wang, W and Li D 2009 Properties of exchange interaction in Yb3Fe5O12 under extreme conditions *Journal of Magnetism and Magnetic Materials* **321** 3307
[15] Wickersheim K A and White R L 1962 Anisotropy of exchange in ytterbium iron garnet *Physical Review Letters* **8** 483
[16] Kimizuka N and Katsura T 1975 The standard free energy of formation of YbFe2O4, Yb2Fe3O7, YbFeO3, and Yb3Fe5O12 at 1200°C *Journal of Solid State Chemistry* **15** 151
[17] Kimizuka N et al 1983 The stability of the phases in the Ln2O3-FeO-Fe2O3 systems which are stable at elevated temperatures (Ln: Lanthanide elements and Y) *Journal of Solid State Chemistry* **49** 65

IOP Conf. Series: Materials Science and Engineering **465** (2019) 012003 doi:10.1088/1757-899X/465/1/012003

[18] Muan A and Osborn E F 1965 *Phase equilibria among oxides in steelmaking* (Addison-Wesley Pub. Co)

The effect of UV aging on structural polymers

L Markovičová and V Zatkalíková

University of Žilina, Faculty of Mechanical Engineering, Department of Material Engineering, Univerzitná 8215/1, 01026 Žilina, Slovakia

lenka.markovicova@fstroj.uniza.sk

Abstract. Polymer composites, their composition, and their properties are the major causes of their application as a construction material. Last century composites with polymer matrix were considered as the most perspective constructional material. Previous researches have only focused on mechanical characteristics of polymer composites but it is also important to observe their mechanical properties before and after UV radiation [1, 2]. The aim of this study is to analyse the change in mechanical properties which are influenced by UV radiation. It was decided that the optimal method for this investigation was to accelerate ageing in UV box. The test samples were evaluated by the selected mechanical parameters - tensile strength, three-point bend, hardness, Vicat temperature. The same parameters were assessed before and after 500 hours UV radiation. These results extend our knowledge of polymer composites and their behaviour before and after UV radiation. This work shown UV radiation has significant influence on the change of selective mechanical properties. Current polymer composites are sensitive to UV radiation which minimize their mechanical properties which change due to partial cross-linking of the matrix and crack at the interface of the matrix and reinforcement.

1. Introduction

Polymeric construction materials are currently developing tremendously and are used in almost all industries. Due to their price, properties, production and processing technologies, more and more classic materials (metals, wood, ceramics, glass, etc.) are replaced. Despite the fact that the use of plastics in various industries is beneficial in terms of energy (their production is less energy-consuming, saving energy resources), environmental (reducing the amount of air pollutants) and economic (saving the raw material resources) we not forget the fact that also polymeric materials become significant waste [3]. Many of them, in degradation processes, release low-molecular compounds that can be harmful to health or endanger the environment. It is therefore important to know the properties of plastics in various degradation environments. In this case we are talking about aging plastics [4].

Polymer aging is the sum of the irreversible changes in the polymer properties of light, sun, ambient air, oxygen, radiation, and heat. Often, many agents act together and the overall effect of aging is more pronounced. Aging is manifested by permanent changes in some properties, in particular loss of ductility, impact strength, sometimes by decreasing strength and loss of polymer mass. In Table 1, there are the basic definitions of terms in the field of aging of polymers [5-6].

Table 1. Selected definitions of the aging of polymers.

Natural aging	Aging of polymers through the environment when stored or in use
Artificial aging	Aging of polymers in artificially created conditions
Climate aging	Aging of polymers due to climatic conditions of a given earth band
Biological aging	Aging of polymers through the action of living organisms that are in contact with them
Mechanical aging	Aging of polymers due to long-term static or dynamic loads
Thermal aging	Aging of polymers through heat
Thermo-oxidative aging	Aging of polymers by the effect of heat in the presence of oxygen
Oxidation aging	Aging of Polymers by Oxygen
Ozone aging	Aging of polymers by ozone
Light aging	Aging of polymers through the visible and ultraviolet portion of the light spectrum
Photooxidation aging	Aging of polymers by the effect of visible and ultraviolet light spectrum in the presence of oxygen
Chemical aging	Aging of polymers by chemically aggressive substances

If polymers are exposed to weathering, we can observe changes in appearance and changes in physical properties after a while. This aging is a result of the relatively long-lasting effect of solar radiation, temperature, oxygen, humidity, ozone, atmospheric impurities. Photochemical destruction and photochemically activated oxidation of macromolecules are considered the main cause of the change. However, the experimental results obtained are very dependent on the climate from which the samples are subjected, the slope of the sample to the sun, the color and the quality of the substrate material on which the samples are laid, the sample preparation conditions, so it is rather difficult to choose a unit to compare the measurement results in samples exposed to weather conditions. A comprehensive approach to solving the problem is to monitor the effect of all factors that are considered in the atmospheric environment to change the physical properties of the polymers individually but also their combinations [7-8].

2. Experimental material and methods

The choice of materials was focused on thermoplastics, which are widely used mainly in the automotive industry. These are in particular composites based on polyamides with filler. As the experimental material on the matrix polyamide 66 (PA 66) was added with polyimide addition. The PA 66 business designation Durethan is intended for injection molding. It is a partially crystalline polymer, characterized by a combination of properties particularly suited for technical applications. High mechanical strength and stiffness is combined with good electro-insulating properties, high thermal and chemical resistance, good weathering and low friction. It dampens shock and noise well. Test samples with the designation PA 66 GF 10, PA 66 GF 20, PA 66 GF 30 differed in glass fiber (GF) content (10 %, 20 % and 30% GF, % = vol. %). The glass fibers coded GF 672 with diameter 10 μm and length 4 mm. The composites are estimated for interior and exterior details of cars, then it is expected their UV stability. The matrix includes also UV stabilizer. The structures of the tested composites were observed by light microscope. We monitored homogeneity of the composite and the manner of distribution of glass fibres in the polymer matrix. The filler content of 10 %, the structure more evenly - glass fibers are evenly distributed in the polymer matrix. Fiber size was also about the same. To 20 % filler content sites began to appear smaller clusters of glass fibers and emptier place in the matrix. Size fiberglass was also maintained at the approximate size. At 30 % filler content the structure of the least uniform. Fiberglass seats were piled on each other. Processing technology will brake the fibers into smaller pieces, so the material was located fibers of different sizes. Different size glass fiber in a polymer matrix was dependent on its percentage (Figure 1) [9].

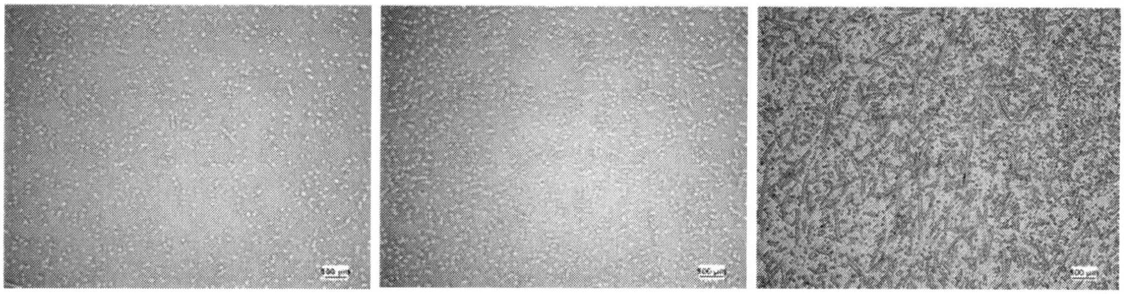

a) 10 % GF b) 20 % GF c) 30 % GF

Figure 1. Structure of composite.

Testing degradation of polymeric materials is one of the most important tests to the lifetime of polymer product. Ageing tests can be either in real conditions of use of the polymer in a particular application, or using artificial accelerated ageing conditions. Accelerated ageing methods provide test results significantly faster natural aging tests. Testing is based on exposing test bars to man-made climate. After a fixed interval of exposure changes are detected in end points (aesthetic, physical, electrical, etc.). The apparatus for man-made weather ageing ensure continued maintaining of artificial climatic conditions (day and night cycles, changing humidity, drought and wet, etc.). The apparatus SolarBox 1500e (Figure 2) which was used in the experiment has the following characteristics: a source of light radiation guarantees a radiant flux of radiation intensity 550 $W.m^{-2}$. The source of light is a xenon arc lamp, but other sources of radiation are allowed too. The device must be equipped with a thermometer built into the black panel, which senses the temperature of the black panel. The black panel temperature of exposure time was selected at 100 °C the liquid phase lasted for 102 minutes and the wet phase for 18 minutes. If it necessary wetting by distilled or deionised water can be applied. Test runs continued for a period fixed in the testing program. The duration of the test was 500 hours [10].

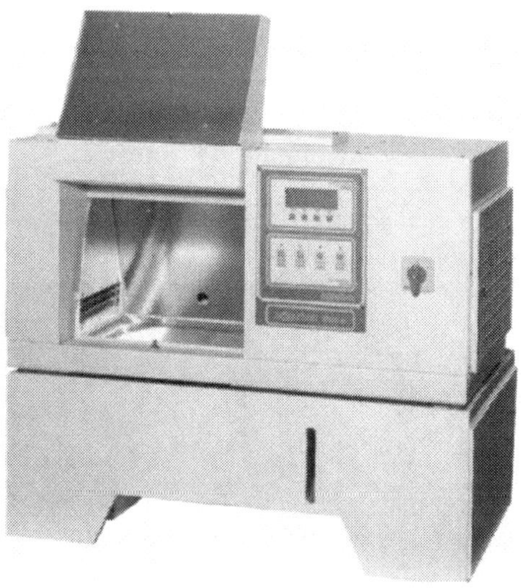

Figure 2. SolarB ox 1500e.

The test samples were evaluated by the selected mechanical parameters (tensile strength, bending strength - three-point bend, hardness - Brinell, Vicat temperature) before and after UV radiation. The parameters were assessed after 1 run, 500 hours of UV radiation acting on the sample.

3. Experimental results and discussion

A tensile test was performed using the device WDW 20. The speed of a moving jaw was set at 1 mm.min^{-1}. The sample was pinned to the jaw and force was exercised in the longitudinal axis until the sample break. We measured the maximum force required to break the test samples. Figure 3 shows that the highest tensile strength of the composite is reached with 30 % glass fiber content on the other hand the smallest tensile strength composite with a filler content of 10 %. It follows that the strength characteristics of the composite depends mainly on the content and distribution of the glass fibers in the polymer matrix. UV radiation is the strength values did not change significantly. Since a partially crystalline PA, stroke causes the chains straighten it, undergoing its subsequent solidification.

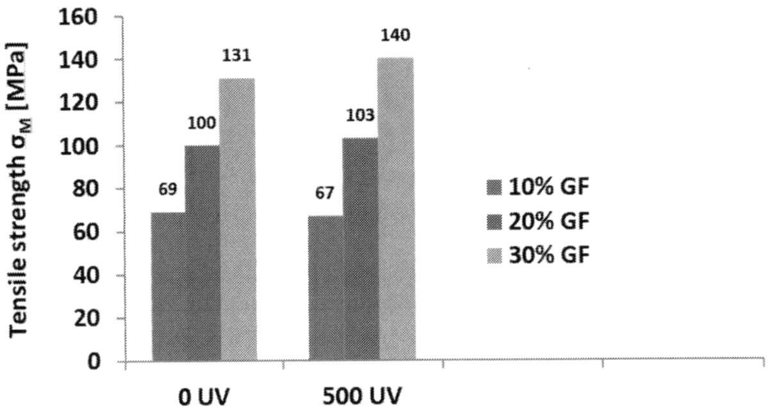

Figure 3. Effect of UV radiation on the change of tensile strength.

A bend test was also performed using the device WDW 20 with the aid of three-point bending. The sample is put on two supports placed, 40 mm away from each other. Burden was applied in the middle of the sample bar to exercise some force. Print speed was also one arm 1 mm.min^{-1}. Peak force was recorded during the test. The test results are indicated in Figure 4.

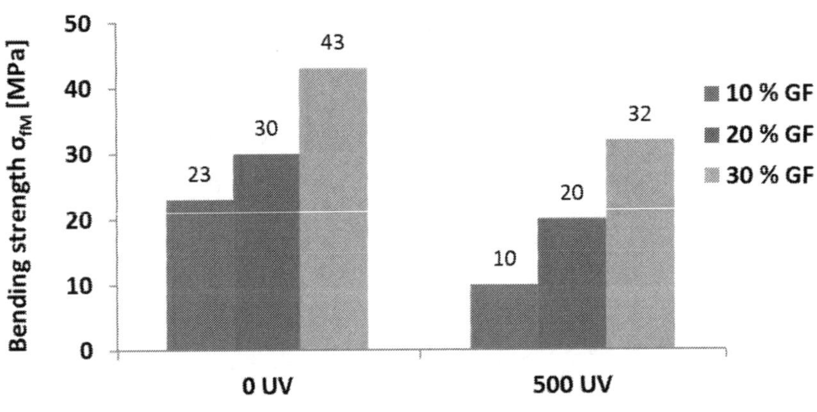

Figure 4. Effect of UV radiation on the change of bending strength.

The highest bending strength of the composite is reached with 30 % glass fiber content on the other hand the smallest bending strength composite with a filler content of 10 %. It follows that the strength characteristics of the composite depends mainly on the content and distribution of the glass fibers in the polymer matrix. Composite specimens showed a significant decrease in the values of the bending strength after 500 hours of UV irradiation. The bending strength decreased by more than 50 % compared to the bending strength that was measured in samples without UV irradiation. Composites are becoming significantly fragile.

Hardness (HBW) test was conducted durometer CV-3000 LDB. Loading force was set to 250 kp, injection time was 10 seconds and the indentation diameter balls of 5 mm. Each sample was made 5 measurements. The final hardness is shown in Figure 5.

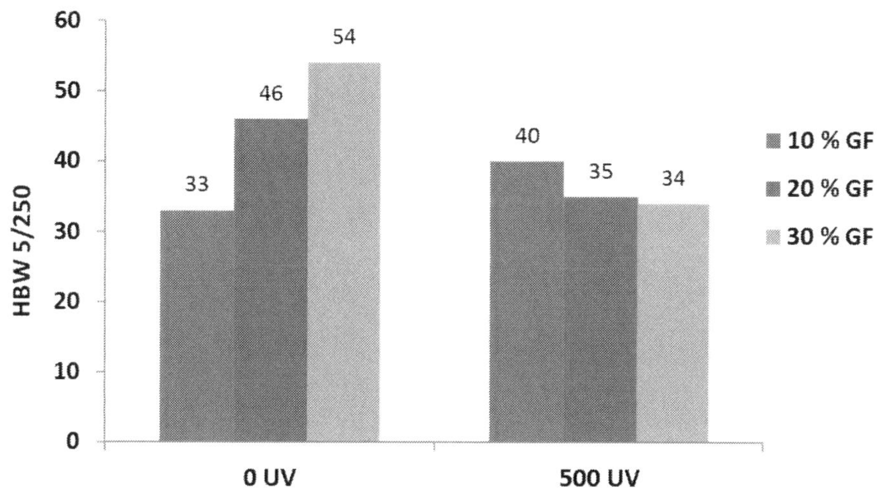

Figure 5 Effect of UV radiation on the change of Brinell hardness

The Vicat softening temperature is the temperature at which a flat-ended needle penetrates the specimen to the depth of 1 mm under a specific load. The temperature reflects the point of softening to be expected when a material is used in an elevated temperature application. The test specimen must be between 3 and 6.5 mm thick and at least 10 mm in width and length. Of each sample were made three measurements and are determined by the final arithmetic value (Table 2).

Table 2. Vicat temperature

	Vicat temperature [°C]	
Filler content	**0 UV**	**500 UV**
10 %	231	234
20 %	240	241
30 %	242	243

4. Conclusions

Based on the experiments performed on composite PA 66 with different content of the filler (GF) we can conclude:

- UV radiation has negligible influence on tensile strength values. Tensile strengths have only slightly decreased after exposure to UV radiation (500 hours).
- The bending strength of the composite depends mainly on the content and distribution of the glass fibers in the polymer matrix. Composite specimens showed a significant decrease

in the values of the bending strength after the 500 hours of UV irradiation. The bending strength decreased by more than 50 % compared to the bending strength that was measured in samples without UV irradiation. Composites are becoming significantly fragile.

- By measuring the Brinell hardness of the material is confirmed that the increasing volume of glass fibers increases the hardness, the UV light had almost no effect on its value.
- The softening point of the composite increases with the content of glass fibers. UV radiation has had no significant impact on softening point, there was only a slight change in the material properties. The biggest change in the softening point was the composite containing 10 % filler wherein the temperature is increased from 231 °C to 243 °C. This may be caused by changes in the structure.
- Materials selected to test mechanical properties are characterized by properties such as stiffness, tensile strength, bending strength, and toughness. Under the influence of the degradation environment, the properties under consideration have changed only minimally, and their use in the consumer and automotive industries has little effect on their utility value.

Acknowledgement
The research was supported partially by Slovak Academy of Science grant KEGA No. 049ŽU-4/2017.

References

[1] Rossi F and Williams G 1997 A New Era in Glass Fiber Composites Proc. 28th AVK Conf. Germany
[2] Miller D M 1987 *Glass Fibers, Composites* vol. 1 Engineering Materials Handbook ASM International
[3] Chua S M and Henderson P J 1991 *Mater Sci Lett* **10** 23 1379–1380
[4] Barry D B and Delatychi O 1992 *Polymer* **33** 6 1261-1265
[5] *Encyclopedia of Materials Science and Technology in Polymer Matrix Composites* 2001 Elsevier Science Ltd ISBN 0-08-0431526
[6] Van der Vegt A K 2002 *From Polymers to Plastic* Delft University Press, Delft, Netherlands.
[7] Mallick P K 2007 *Fiber-Reinforced Composites, Materials, Manufacturing, and Design*, 3rd Ed. CRC Press, Taylor & Francis Group ISBN-13: 978-0-8493-4205-9
[8] Raz K, Zahalka M and Chval Z 2017 *Manufacturing Technology* **17** 79-83
[9] Markovicova L, Zatkalikova V and Vasko A 2016 *Manufacturing Technology* **6** 1033-1037
[10] Standards STN EN ISO 4892 Plastics – Methods of Exposure to Laboratory Light Sources.

IOP Conf. Series: Materials Science and Engineering **465** (2019) 012005 doi:10.1088/1757-899X/465/1/012005

An Estimating the Local Deformation Parameters in Grained Structures via Stereological Methods

S Minárik[1] and M Martinkovič[2]
[1] Slovak University of Technology in Bratislava, Faculty of Materials Science and Technology in Trnava, Advanced Technologies Research Institute, Ulica Jána Bottu č. 2781/25, 917 24 Trnava, Slovakia
[2] Slovak University of Technology in Bratislava, Faculty of Materials Science and Technology in Trnava, Institute of Production Technologies, Ulica Jána Bottu č. 2781/25, 917 24 Trnava, Slovakia

stanislav.minarik@stuba.sk

Abstract. A lot of sophisticated methods have been developed to study grained structures by 2D imaging such as image correlation method, triple points technique, electron backscatter diffraction (EBSD), computer tomography (CT), etc. With applying 3D volume data mentioned methods can significantly help in precise investigation of grained structures. However these methods have several limitations. They are neither always economically efficient nor can be applied for different situations. In this paper we present relationship between change of grain`s shape caused by deformation and deformation parameters. Next we investigate a ratio ξ between total surface area of images of all grains in structure displayed in orthogonal 2D projection and their total surface area in the 3D (that means entire surface of grain boundaries). During an applied mechanical loading the grained structure is deformed and this ratio ξ varies. One specific conclusion of our work is that in cases of giant random grained structures mentioned ratio is a definite smooth function of deformation parameters. This result suggests that deformation of grained structure can be intuitively estimated on the basis of grains surface distortion in given area of bulk sample which is visible on metallographic cut. Unfortunately 2D images of real grains in orthogonal projection overlap each other and stereological methods must be used in precise deformation analysis.

1. Introduction

If the loading in grained structure spatially varies it is difficult to determine the value of the local strain experimentally. Local strain distribution can be only intuitively estimated on the basis of changes of grains spatial orientation caused by structure deformation. Grains orientation can be easily evaluated experimentally by stereological methods.

Stereology provides the three-dimensional interpretation of two-dimensional cross sections of materials. Stereological techniques allow the extracting of quantitative information about a three-dimensional material from measurements made on its two-dimensional planar sections. Despite the fact that stereological methods are quite old indeed nothing new and better has been developed in this area so far. These methods are widely used today especially in the investigation of the character of grained and foamed structures.

Spatial orientation of grained structure can be experimentally evaluated by statistical Saltykov method of oriented lines [1-2] directly from metallographic cut of bulk sample. If the correlation between the change of grains spatial orientation and deformation parameters would be known a detailed deformation map in the structure could be designed. Our work is motivated by the fact that exact theoretical basis for assessment of deformation by means of the change of orientation is still missing. We suggest that results of such measurement, i.e. degree of grains orientation in volume of plastically deformed structure, can be converted to values of local deformation parameters.

Presented paper contains a mathematical model for describing the deformation of a fictional, simplified grained structure. No real material is specified, nor is the actual deformation process described. We want to contribute to the theoretical work on this topic published in the past [3-6]. Purely theoretical approach is presented. However it have been shown that such a deformation mechanism, i.e. elongation (or the shortening) of grains under applied load, exists in a real-world material [7-8].

2. Grain-surface distortion at linear deformation

During an applied mechanical loading, according to the different orientations and the influences from neighboring grains, most of grains are submitted to a deformation. The deformation of grain results in a change in orientation and amount of grain surface area. This is evident in the case if the typical grain prolongation in direction of loading force is visible (see Figure 1. and Figure 2.).

Consider isotropic, linear elastic material with the grained structure. Generally the parametrization of any single grain surface in 3D can be accomplished by spatial coordinates of each points on the surface:

$$X_i = X_i(u,v), \text{ where } i = 1, 2, 3 \tag{1.1}$$

and u, v are parameters that changes in appropriate intervals. Therefore the position vectors of each points on the surface of grain with any possible shape in undeformed cofigurations can be written as:

$$\vec{R}(u,v) = \left[X_1(u,v), X_2(u,v), X_3(u,v)\right] \quad , \quad u \in \langle u_1, u_2 \rangle \quad v \in \langle v_1, v_2 \rangle . \tag{1.2}$$

Let the position vectors of each points on the surface of the same grain in deformed cofigurations are

$$\vec{r}(u,v) = \left[x_1(u,v), x_2(u,v), x_3(u,v)\right] \quad , \quad u \in \langle u_1', u_2' \rangle \quad , \quad v \in \langle v_1', v_2' \rangle . \tag{1.3}$$

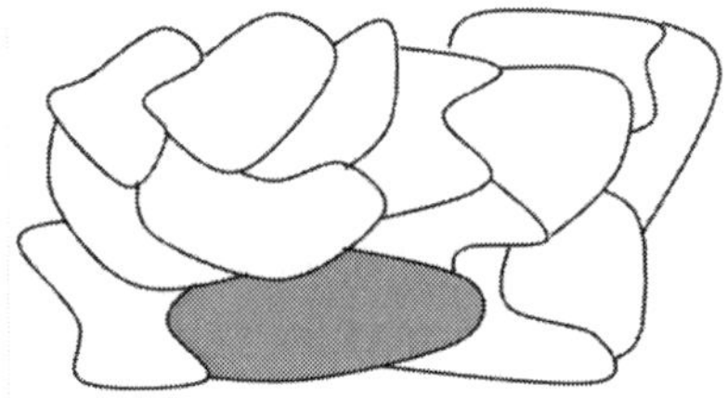

Figure 1. Scheme illustrating a view on an undeformed grained structure cross-section (for example on metallographic cut). Scheme shows cross-section of some individual grains.

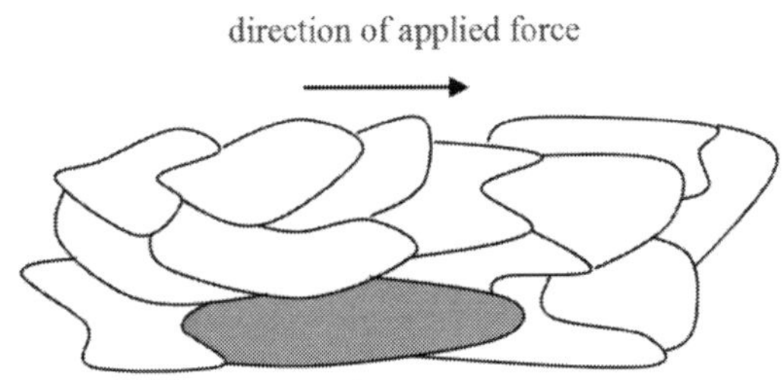

Figure 2. Scheme illustrating a view on the same grained structure cross-section in deformed configuration. Scheme demonstrates grain surface area distortion and typical prolongation of individual grains.

In general case, any 3D deformation can be described by a 3x3 deformation tensor

$$\vec{\vec{\varepsilon}} = \begin{pmatrix} \varepsilon_{11}, & \varepsilon_{12}, & \varepsilon_{13} \\ \varepsilon_{21}, & \varepsilon_{22}, & \varepsilon_{23} \\ \varepsilon_{31}, & \varepsilon_{32}, & \varepsilon_{33} \end{pmatrix} \tag{1.4}$$

Using notation mentioned above the grain linear deformation (for small strain) can be expressed as anisomorphic linear transformation

$$\vec{r} = \vec{\vec{\varepsilon}} \cdot \vec{R}, \quad \text{or} \quad x_i = \varepsilon_{ij} X_j, \tag{1.5}$$

where $i, j = 1, 2, 3$ and ε_{ij} are coefficients of deformation tensor (1.4). It is no problem to verify that $\varepsilon_{ij} = \delta_{ij}$ in undeformed case (where δ_{ij} is Kronecker symbol). We used typical Einstein`s summation convention in the second formula in (1.5) whereby when an index variable appears twice in a single term it implies summation of that term over all the values of the index. For the simplicity this convention is used in all next text.

Vector of a surface element of undeformed grain is determined as follows

$$d^2\vec{S}_{undefor} = \left(\frac{\partial X_2}{\partial u}\frac{\partial X_3}{\partial v} - \frac{\partial X_3}{\partial u}\frac{\partial X_2}{\partial v}\right)\vec{i} - \left(\frac{\partial X_1}{\partial u}\frac{\partial X_3}{\partial v} - \frac{\partial X_3}{\partial u}\frac{\partial X_1}{\partial v}\right)\vec{j} + \left(\frac{\partial X_1}{\partial u}\frac{\partial X_2}{\partial v} - \frac{\partial X_2}{\partial u}\frac{\partial X_1}{\partial v}\right)\vec{k} \quad (1.6)$$

An example of this vector is shown in the Figure 3. Using transformation (1.5) the vector of same surface element of the grain in deformed configuration can be written in the form

$$d^2\vec{S}_{defor} = \varepsilon_{2j}\varepsilon_{3k}\left(\frac{\partial X_j}{\partial u}\frac{\partial X_k}{\partial v} - \frac{\partial X_j}{\partial v}\frac{\partial X_k}{\partial u}\right)\vec{i} - \varepsilon_{1j}\varepsilon_{3k}\left(\frac{\partial X_j}{\partial u}\frac{\partial X_k}{\partial v} - \frac{\partial X_j}{\partial v}\frac{\partial X_k}{\partial u}\right)\vec{j} + \varepsilon_{1j}\varepsilon_{2k}\left(\frac{\partial X_j}{\partial u}\frac{\partial X_k}{\partial v} - \frac{\partial X_j}{\partial v}\frac{\partial X_k}{\partial u}\right)\vec{k}, \quad (1.7)$$

where $i, j, k = 1, 2, 3$. This vector is illustrated in the Figure 4. After suitable treatment and re-indexing of sums in formula (1.7) the change of mentioned vector of surface element during deformation can be expressed as

$$d^2\vec{S}_{defor} = \vec{\vec{\sigma}} \cdot d^2\vec{S}_{undefor} \quad (1.8)$$

where $\vec{\vec{\sigma}}$ can be defined as a *tensor of grain surface deformation*. Coefficients of this tensor are related to coefficients of deformation tensor.

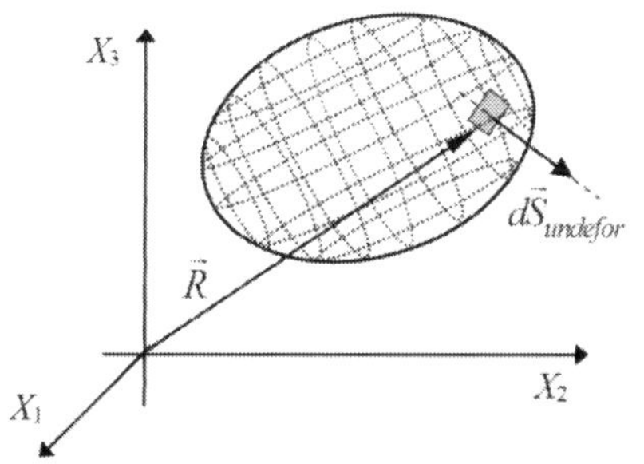

Figure 3. Undeformed grain surface.

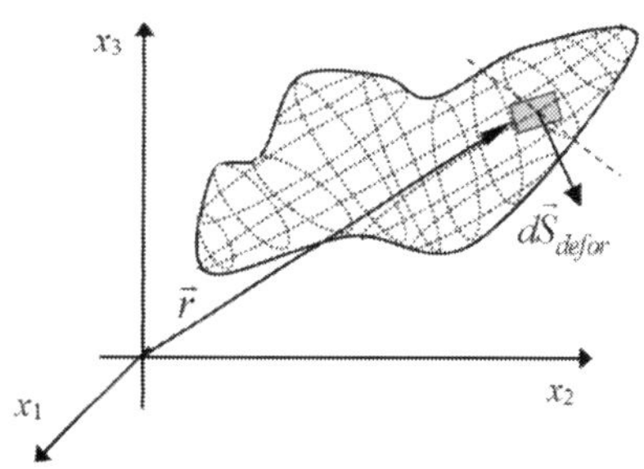

Figure 4. Deformed grain surface.

$$\sigma_{wm} = (-1)^{w+m}\, det\left(\vec{\vec{\varepsilon}}_{mw}\right)$$

where $det\,(\vec{\vec{\varepsilon}}_{mw})$ is the minor (or subdeterminant) of matrix of the deformation tensor $\vec{\vec{\varepsilon}}$. This minor is obtained from the matrix of tensor $\vec{\vec{\varepsilon}}$ by the deleting of m-th row and w-th column. So it is easily to show that coefficients of tensor of grain surface deformation can be found as

$$\sigma_{wm} = \varepsilon_{ik}\varepsilon_{jn} - \varepsilon_{in}\varepsilon_{jk} \quad (1.9)$$

where indexes

$$i = \frac{9 - 2w - (-1)^w}{4},\ j = \frac{15 - 2w + (-1)^w}{4}, \quad (1.10)$$

$$k = \frac{12 - 2m - \{3 + (-1)^m\}(-1)^{m+w}}{4},\ n = \frac{12 - 2m + \{3 + (-1)^m\}(-1)^{m+w}}{4},\ w, m = 1, 2, 3. \quad (1.11)$$

Then the magnitude of vector (1.8), i.e. size of the deformed surface element is given by

$$\left(d^2 S_{defor}\right)^2 = \Gamma_{jnks} \frac{\partial X_j}{\partial u} \frac{\partial X_n}{\partial u} \frac{\partial X_k}{\partial v} \frac{\partial X_s}{\partial v}, \tag{1.12}$$

where coefficients

$$\Gamma_{jnks} = \sum_{\substack{V(\{p,q\}),p\neq q \\ p,q\in\{1,2,3\}}} \varepsilon_{pj}\varepsilon_{qk}\left(\varepsilon_{pn}\varepsilon_{qs} - \varepsilon_{qn}\varepsilon_{ps}\right) . \tag{1.13}$$

Notation under the sumation symbol in (1.13) means sum over all variations of indexes p, q while $p \neq q$. The entire surface area of the deformed grains can be determined by formula

$$S_{defor} = \int_{v_1}^{v_2}\int_{u_1}^{u_2} \sqrt{\Gamma_{jklm} \frac{\partial X_j}{\partial u} \frac{\partial X_k}{\partial v} \frac{\partial X_l}{\partial v} \frac{\partial X_m}{\partial u}} dudv . \tag{1.14}$$

where $j, k, l, m = 1, 2, 3$. We note, that we permanently use Einstein`s sumation convention and the sum of 81 members is under the square root in formula (1.14).

3. Grain surface 2D projection

Orientation of each grain-surface elements in 3D is determined by vector $d\vec{S}$ which is perpendicular on the surface. Planar character of infinitesimal surface area enables clearly define orientation of surface element in 3D just by means of the vector $d\vec{S}$. Hence it is possible to analyze change of surface element spatial orientation during grain deformation. Parameter ξ defined by means of orthogonal 2D projection of the element onto the plane characterized by unit vector $\vec{\tau}$ ca be used:

$$d\xi = \frac{\left|\vec{\tau}\cdot d\vec{S}\right|}{S} , \tag{2.1}$$

where S is surface area size of whole grain. Absolute value in the numerator determines the size of the orthogonal projection of the surface element dS onto the plane perpendicular to the vector $\vec{\tau}$.
By integrating of (2.1) over the entire grain surface we get:

$$\xi(\vec{\tau}) = \frac{1}{S} \int_{(Surface)} \left|\vec{\tau}\cdot d\vec{S}\right| . \tag{2.2}$$

Parameter ξ in both deformed and undeformed configurations can be quantified by (2.2):

$$\xi_{undefor} = \frac{\int_{(surface)} \left|\vec{\tau}\cdot d\vec{S}_{undefor}\right|}{\int_{(surface)} \left|d\vec{S}_{undefor}\right|} \quad (a) \quad , \quad \xi_{defor} = \frac{\int_{(surface)} \left|\vec{\tau}\cdot d\vec{S}_{defor}\right|}{\int_{(surface)} \left|d\vec{S}_{defor}\right|} \quad (b) . \tag{2.3a,b}$$

Let $\vec{\rho}$ is unit vector paralel to vector of grain-surface element of undeformed grain, i.e.

$$d\vec{S}_{undefor} = \vec{\rho} dS_{undefor} \quad \text{where } \left|\vec{\rho}\right| = 1 .$$

Taking into account the formula (1.8) and considering the fact:

$$\vec{\tau}\cdot d^2\vec{S}_{defor} = \vec{\tau}\cdot\left(\vec{\vec{\sigma}}\cdot d^2\vec{S}_{undefor}\right) = \left(\vec{\vec{\sigma}}^T\cdot\vec{\tau}\right)\cdot d^2\vec{S}_{undefor}, \tag{2.4}$$

where $\vec{\vec{\sigma}}^T$ is transposed tensor to the grain surface deformation tensor $\vec{\vec{\sigma}}$ determined by (1.9), formula (2.2) in deformed configuration can be rewritten to the next form

$$\xi_{defor} = \frac{\int\limits_{(Surface)} \left| \left(\vec{\sigma}^T \cdot \vec{\tau} \right) \cdot d\vec{S}_{undefor} \right|}{\int\limits_{(Surface)} \left| \vec{\sigma} \cdot \vec{\rho} \right| dS_{undefor}} . \qquad (2.5)$$

As can be concluded from comparison of above equations (2.3a) and (2.5), transformation of parameter ξ during grain deformation can be realized as follows:

$$\vec{\tau} \rightarrow \vec{\sigma}^T \cdot \vec{\tau} \;\; \wedge \;\; \vec{\rho} \rightarrow \vec{\sigma} \cdot \vec{\rho} \; . \qquad (2.6)$$

It can be seen from (2.5) that in general the transformation of ξ during grain deformation appears to be relatively complicated and we will not examine its mathematical properties in detail. However, it should be noted that we integrate over the surface of undeformed grain in both formulas (2.3a) and (2.5). So to detemine the value of parameter ξ in deformed configuration we need to know only the shape of undeformed grain and coefficients of deformation tensor ε_{ij}. Therefore we believe, that ξ is suitable for the quantitative evaluation of grain-surface distortion during local deformation of grained structure. (1.14).

4. Random matrix approximation

Integrals in formula (2.5) can be modeled by summation over infinitely small areas placed on grains boundaries (see Figure 5.). So we get

$$\xi_{defor} \approx \frac{\sum\limits_{i=1}^{N} \left| \left(\vec{\sigma}^T \cdot \vec{\tau} \right) \cdot \vec{S}^{(i)}_{undefor} \right|}{\sum\limits_{i=1}^{N} \left| \vec{\sigma} \cdot \vec{\rho}^{(i)} \right| S^{(i)}} \qquad (3.1)$$

where $\vec{\tau} = [\tau_1, \tau_2, \tau_3]$ is unit vector perpendicular to the the orthogonal projection plane ($\tau_1^2 + \tau_2^2 + \tau_3^2 = 1$) and $\vec{\rho}^{(i)}$ is unit vector perpendicular to the i-th grain-surface element of undeformed grain

$$\vec{S}^{(i)}_{undefor} = \vec{\rho}^{(i)} S^{(i)} . \qquad (3.2)$$

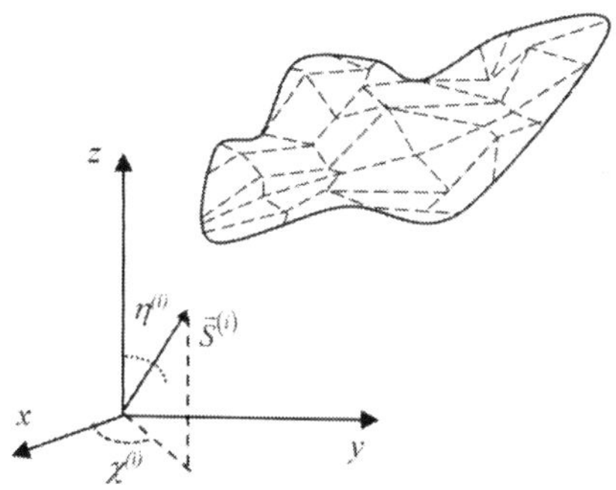

Figure 5. Scheme illustrating a view on the model of the single grain.

Vector $\vec{\rho}^{(i)}$ can be written in the form

$$\vec{\rho}^{(i)} = \left[sin\left(\eta^{(i)}\right) cos\left(\chi^{(i)}\right), sin\left(\eta^{(i)}\right) sin\left(\chi^{(i)}\right), cos\left(\eta^{(i)}\right) \right]$$

while angles $\eta^{(i)}$ and $\chi^{(i)}$ determines orientation of the i-th grain-surface element in 3D and then

$$\vec{S}^{(i)}_{undefor} = \left[S^{(i)}_1, S^{(i)}_2, S^{(i)}_3 \right] = \left[S^{(i)} sin\left(\eta^{(i)}\right) cos\left(\chi^{(i)}\right), S^{(i)} sin\left(\eta^{(i)}\right) sin\left(\chi^{(i)}\right), S^{(i)} cos\left(\eta^{(i)}\right) \right] \qquad (3.3)$$

If we use result (1.9) and notations (3.3), the formula (3.1) can be calculated as

$$\xi_{defor} \approx \frac{\sum\limits_{i=1}^{N} \left(S^{(i)} \left| \sum\limits_{n=1}^{3} \tau_n \, det \, \tilde{A}^{(i)}_n \right| \right)}{\sum\limits_{i=1}^{N} \left(S^{(i)} \sqrt{\sum\limits_{n=1}^{3} \left(det \, \tilde{A}^{(i)}_n \right)^2} \right)} \quad \ldots \; n = 1, 2, 3 \qquad (3.4)$$

where $\tilde{A}^{(i)}_n$ are following matrix-valued variables

$$\vec{A}_1^{(i)} = \begin{pmatrix} \sin\left(\eta^{(i)}\right)\cos\left(\chi^{(i)}\right), & \sin\left(\eta^{(i)}\right)\sin\left(\chi^{(i)}\right), & \cos\left(\eta^{(i)}\right) \\ \varepsilon_{21}, & \varepsilon_{22}, & \varepsilon_{23} \\ \varepsilon_{31}, & \varepsilon_{32}, & \varepsilon_{33} \end{pmatrix} \tag{3.5}$$

$$\vec{A}_2^{(i)} = \begin{pmatrix} \varepsilon_{11}, & \varepsilon_{12}, & \varepsilon_{13} \\ \sin\left(\eta^{(i)}\right)\cos\left(\chi^{(i)}\right), & \sin\left(\eta^{(i)}\right)\sin\left(\chi^{(i)}\right), & \cos\left(\eta^{(i)}\right) \\ \varepsilon_{31}, & \varepsilon_{32}, & \varepsilon_{33} \end{pmatrix} \tag{3.6}$$

$$\vec{A}_3^{(i)} = \begin{pmatrix} \varepsilon_{11}, & \varepsilon_{12}, & \varepsilon_{13} \\ \varepsilon_{21}, & \varepsilon_{22}, & \varepsilon_{23} \\ \sin\left(\eta^{(i)}\right)\cos\left(\chi^{(i)}\right), & \sin\left(\eta^{(i)}\right)\sin\left(\chi^{(i)}\right), & \cos\left(\eta^{(i)}\right) \end{pmatrix}. \tag{3.7}$$

Parameter ξ of whole deformed grained structure is calculated by means of the formula (3.4) while sums in the numerator and denominator on the right side of this formula can be determined using (3.5), (3.6) and (3.7).

Any randomly oriented surface of grain boundary $S^{(i)}$ in the given bulk sample with grained structure can be included into the calculation of ξ if corresponding parameters $\eta^{(i)}$ and $\chi^{(i)}$ will be applied. Therefore random values of angles $\eta^{(i)}$ and $\chi^{(i)}$ should be substituted step by step into formula (3.4) when the calculation progress. The problem of parameter ξ calculation can be formulated as a random matrix problems because matrices (3.5), (3.6) and (3.7) can be smartly used in the formal mathematical notation as well as numerical implementation of mentioned approximation approach.

5. The case of plastic deformation

Fundamental assumption used to establish the theory of plasticity is that plastic deformation is isochoric or volume preserving. Next we apply results obtained in previous section 3 in the case of plastic grain deformation in which the change in volume of grain is zero.

Grain volume conservation during plastic deformation is given by formula

$$V_{undefor} = V_{defor}\,. \tag{4.1}$$

Gauss-Ostrogradski integral theorem can be used for determination of grain volume in both undeformed ($V_{undefor}$) and deformed (V_{defor}) configurations

$$\int\limits_{(S_{undefor})} \vec{R}\cdot d\vec{S}_{undefor} = \int\limits_{(V_{undefor})} div\left(\vec{R}\right)dV = 3V_{undefor}\,, \quad \int\limits_{(S_{defor})} \vec{r}\cdot d\vec{S}_{defor} = \int\limits_{(V_{defor})} div\left(\vec{r}\right)dV' = 3V_{defor}\,. \tag{4.2}$$

Next formula results from the comparison of equations (4.2)

$$\int\limits_{(S_{undefor})} \vec{R}\cdot d\vec{S}_{undefor} = \int\limits_{(S_{defor})} \vec{r}\cdot d\vec{S}_{defor} = \int\limits_{(S_{undefor})} \left\{\vec{\vec{\sigma}}^T\cdot\left(\vec{\vec{\varepsilon}}\cdot\vec{R}\right)\right\}\cdot d\vec{S}_{undefor}\,, \tag{4.3}$$

where we used (1.5) and (1.8). We integrate over the same (but any) area on the both sides of previous equation (4.3). Therefore for plastic deformation apply

$$\vec{R} = \vec{\vec{\sigma}}^T\cdot\left(\vec{\vec{\varepsilon}}\cdot\vec{R}\right), \qquad \vec{\vec{\sigma}}^T = \vec{\vec{\varepsilon}}^{-1} \text{ respectively.} \tag{4.4}$$

Transposed tensor to the grain surface deformation tensor is inverse to tensor of deformation in this case. If we consider (4.3), next equation can be written in coordinate form

$$\left\{\left(\varepsilon_{1n}\varepsilon_{2j}\varepsilon_{3k}-\varepsilon_{2n}\varepsilon_{1j}\varepsilon_{3k}+\varepsilon_{3n}\varepsilon_{1j}\varepsilon_{2k}\right)-\left(\delta_{1n}\delta_{2j}\delta_{3k}-\delta_{2n}\delta_{1j}\delta_{3k}+\delta_{3n}\delta_{1j}\delta_{2k}\right)\right\}\left(\frac{\partial X_j}{\partial u}\frac{\partial X_k}{\partial v}-\frac{\partial X_j}{\partial v}\frac{\partial X_k}{\partial u}\right)X_n=0\,. \quad (4.5)$$

Equation (4.5) must be satisfied for any function $X_i = X_i(u,v)$ when the grain is plastically deformed. If (4.5) is expanded without further assumptions, it will lead to the following 27 equations

$$\begin{aligned}&\varepsilon_{1n}\left(\varepsilon_{2j}\varepsilon_{3k}-\varepsilon_{2k}\varepsilon_{3j}\right)-\varepsilon_{2n}\left(\varepsilon_{1j}\varepsilon_{3k}-\varepsilon_{1k}\varepsilon_{3j}\right)+\varepsilon_{3n}\left(\varepsilon_{1j}\varepsilon_{2k}-\varepsilon_{1k}\varepsilon_{2j}\right)-\\&\quad-\delta_{1n}\left(\delta_{2j}\delta_{3k}-\delta_{2k}\delta_{3j}\right)+\delta_{2n}\left(\delta_{1j}\delta_{3k}-\delta_{1k}\delta_{3j}\right)-\delta_{3n}\left(\delta_{1j}\delta_{2k}-\delta_{1k}\delta_{2j}\right)=0 \quad \ldots\forall \quad n,j,k \qquad (4.6)\end{aligned}$$

i.e. next equation must be satisfied in case of plastic deformation

$$det\left(\vec{\vec{\varepsilon}}\right)=1\,. \qquad (4.7)$$

If the homogeneous and isotropic grained structure is considered, following conditions for the deformation tensor can be applied

$$\varepsilon_{ji}=0 \text{ for all } i\neq j\neq k \quad \wedge \quad det\left(\vec{\vec{\varepsilon}}\right)=\varepsilon_{ii}\varepsilon_{jj}\varepsilon_{kk}=1\,. \qquad (4.8)$$

In the case of plastic deformation of isotropic and homogeneous grained structure tensor elements are related to grain prolongation (contraction) δ ($-1<\delta<\infty$). If the loading force is applied in direction of x-axis it holds

$$\varepsilon_{11}=\delta+1\,,\ \varepsilon_{22}=\varepsilon_{33}=\frac{1}{\sqrt{\delta+1}}\,. \qquad (4.9)$$

Application of (4.8) and (4.9) in formulas (2.4) and (4.4) will simplify integration in (2.5). Nevertheless detail information about the shape of grain is necessary for the practical application because we integrate over entire surface of the grain in this formula. However it is absolutely impossible to determine real shape of each grain in material structure exactly. Therefore deformation of modeled grained structure consisting from grains with various idealized shapes are investigated usually [9].

6. Monte-Carlo simulation of $\xi(\delta)$ for plastically deformed grained structures

In the frame of random matrix approximation presented in the previous section, grain boundaries are considered to be a large set of planar surfaces and therefore all derivatives with respect to parameters u and v in formulas (1.7) and (1.14) are constants. These constants are different for each planar surface and integrals can be replaced by sums. Random variables $\eta^{(i)}$, $\chi^{(i)}$ and $S^{(i)}$ actually represent orientation and size of i – th planar surface in the structure.

We consider random homogeneous grained structure in which the single grain may be of any shape and can have any orientation. The size and orientation of mentioned small flat surfaces in the structure are random in the approximation, they are not limited by particular shape of the grain.

In the case of homogeneous and isotropic plastically deformed grained structure formulas (4.8) and (4.9) must be considered and numerator and denominator on the right side of formula (3.1) can be simplified to the following form

$$\sum_{i=1}^{N}\left|\left(\vec{\sigma}^{T}\cdot\vec{\tau}\right)\cdot\vec{s}^{(i)}\right|=\sum_{i=1}^{N}S^{(i)}\left|\frac{1}{(\delta+1)}sin\left(\eta^{(i)}\right)cos\left(\chi^{(i)}\right)\tau_1+\sqrt{\delta+1}\,sin\left(\eta^{(i)}\right)sin\left(\chi^{(i)}\right)\tau_2+\sqrt{\delta+1}\,cos\left(\eta^{(i)}\right)\tau_3\right| \qquad (5.1)$$

$$\sum_{i=1}^{N}\left|\vec{\sigma}\cdot\vec{\rho}^{(i)}\right|S^{(i)}=\sum_{i=1}^{N}S^{(i)}\sqrt{\left\{\frac{1}{(\delta+1)}cos\left(\chi^{(i)}\right)sin\left(\eta^{(i)}\right)\right\}^2+\left\{\sqrt{\delta+1}\,sin\left(\chi^{(i)}\right)sin\left(\eta^{(i)}\right)\right\}^2+\left\{\sqrt{\delta+1}\,cos\left(\eta^{(i)}\right)\right\}^2}\,. \quad (5.2)$$

where N is number of random variables (i.e. number of considered planar grain boundaries in the model of grained structure).

a) The case of „out of plane" deformation
If the vector $\vec{\tau} = [1,0,0]$ (out of plane deformation), we get next formula by means of substituting (5.1) and (5.2) into (3.1)

$$\xi_{defor} = \frac{\sum_{i=1}^{N} S^{(i)} \left| sin\left(\eta^{(i)}\right) cos\left(\chi^{(i)}\right) \right|}{\sum_{i=1}^{N} S^{(i)} \sqrt{cos^2\left(\chi^{(i)}\right) sin^2\left(\eta^{(i)}\right) + (\delta+1)^3 \left\{ sin^2\left(\chi^{(i)}\right) sin^2\left(\eta^{(i)}\right) + cos^2\left(\eta^{(i)}\right) \right\}}} \quad . \tag{5.3}$$

Parameter ξ calculated via the vector oriented parallel to the direction of deformation force can be evaluated in this simple case by means of formula (5.3). Values $\eta^{(i)}$, $\chi^{(i)}$ and $S^{(i)}$ must be randomly generated and substituted to formula (5.3) during the numerical simulation of the function $\xi_{defor}(\delta)$. However it is also necessary to take into account the value $\xi_{undefor}$ for the grained structure in the undeformed configuration during the simulation. Prolongation $\delta = 0$ in the mentioned undeformed case and then from (5.3) it holds

$$\xi_{undefor} = \xi_{defor}(\delta = 0) = \frac{\sum_{i=1}^{N} S^{(i)} \left| sin\left(\eta^{(i)}\right) cos\left(\chi^{(i)}\right) \right|}{\sum_{i=1}^{N} S^{(i)}} \quad . \tag{5.4}$$

Random values $\eta^{(i)}$, $\chi^{(i)}$ and $S^{(i)}$ must be generated in a such way to meet an initial condition (5.4). Considering (5.3) and (5.4) we get

$$\xi_{defor} = \frac{\xi_{undefor} \sum_{i=1}^{N} S^{(i)}}{\sum_{i=1}^{N} S^{(i)} \sqrt{cos^2\left(\chi^{(i)}\right) sin^2\left(\eta^{(i)}\right) + (\delta+1)^3 \left\{ sin^2\left(\chi^{(i)}\right) sin^2\left(\eta^{(i)}\right) + cos^2\left(\eta^{(i)}\right) \right\}}} \quad . \tag{5.5}$$

We propose to evaluate the value ξ for the grained structure by the formula (5.5) in cases when an extension of this stucture caused by the deformation force, i.e if $0 < \delta$ is observed .

b) The case of „in plane" deformation
In the case if there is contraction of the structure under the deformation force, i.e. $-1 < \delta < 0$ (in plane deformation), it is appropriate to evaluate ξ via any vector oriented perpendicular to deformation force. Vector

$$\vec{\tau} = \left[0, cos(\theta), sin(\theta)\right] \tag{5.6}$$

where $0 < \theta < 2\pi$ must be considered in this case of structure squeezing and numerator and denominator on the right side of formula (3.1) take the form:

$$\sum_{i=1}^{N} \left| \left(\bar{\bar{\sigma}}^T \cdot \vec{\tau}\right) \cdot \vec{\rho}^{(i)} S^{(i)} \right| = \sum_{i=1}^{N} S^{(i)} \left| \sqrt{\delta+1}\, sin\left(\eta^{(i)}\right) sin\left(\chi^{(i)}\right) cos(\theta) + \sqrt{\delta+1}\, cos\left(\eta^{(i)}\right) sin(\theta) \right| \tag{5.7}$$

$$\sum_{i=1}^{N} \left| \bar{\bar{\sigma}} \cdot \vec{\rho}^{(i)} \right| S^{(i)} = \sum_{i=1}^{N} S^{(i)} \sqrt{ \left\{ \frac{1}{(\delta+1)} cos\left(\chi^{(i)}\right) sin\left(\eta^{(i)}\right) \right\}^2 + \left\{ \sqrt{\delta+1}\, sin\left(\chi^{(i)}\right) sin\left(\eta^{(i)}\right) \right\}^2 + \left\{ \sqrt{\delta+1}\, cos\left(\eta^{(i)}\right) \right\}^2 } \tag{5.8}$$

Applying (5.6), (5.7) and (5.8) in formula (3.1) we get

$$\xi'_{defor} = \frac{\sum_{i=1}^{N} S^{(i)} \left| \sqrt{\delta+1}\, sin\left(\eta^{(i)}\right) sin\left(\chi^{(i)}\right) cos\left(\theta^{(i)}\right) + \sqrt{\delta+1}\, cos\left(\eta^{(i)}\right) sin\left(\theta^{(i)}\right) \right|}{\sum_{i=1}^{N} S^{(i)} \sqrt{\left\{\frac{1}{(\delta+1)} cos\left(\chi^{(i)}\right) sin\left(\eta^{(i)}\right)\right\}^2 + \left\{\sqrt{\delta+1}\, sin\left(\chi^{(i)}\right) sin\left(\eta^{(i)}\right)\right\}^2 + \left\{\sqrt{\delta+1}\, cos\left(\eta^{(i)}\right)\right\}^2}} . \quad (5.9)$$

Therefore in undeformed configuration:

$$\xi'_{undefor} = \xi'_{defor}(\delta = 0) = \frac{\sum_{i=1}^{N} S^{(i)} \left| sin\left(\eta^{(i)}\right) sin\left(\chi^{(i)}\right) cos(\theta) + cos\left(\eta^{(i)}\right) sin(\theta) \right|}{\sum_{i=1}^{N} S^{(i)}} \quad (5.10)$$

and we get next formula for the calculation of the ξ in the same way as before:

$$\xi'_{defor} = \frac{\sqrt{\delta+1}\,(\delta+1)\,\xi'_{undefor} \sum_{i=1}^{N} S^{(i)}}{\sum_{i=1}^{N} S^{(i)} \sqrt{cos^2\left(\chi^{(i)}\right) sin^2\left(\eta^{(i)}\right) + (\delta+1)^3 \left\{sin^2\left(\chi^{(i)}\right) sin^2\left(\eta^{(i)}\right) + cos^2\left(\eta^{(i)}\right)\right\}}} . \quad (5.11)$$

Functions $\xi_{defor}(\delta)$ were found using formulas (5.5) and (5.11) numerically by Monte Carlo simulation. Procedure of this simulation is illustrated by means of graphs shown in the Figure 6. (for zero boundary conditions $1\text{-}2\xi_{undef} = 0$). Mentioned graphs were obtain by successive substitution of N randomly chosen triples of variables $\{\eta^{(i)}, \chi^{(i)}, S^{(i)}\}$ into formulas (5.5) and (5.11). Values $(1\text{-}2\xi)$ are displayed on vertical axes in mentioned graphs as these values are assumed to be corresponding to parameters describing spatial orientation of grains from Saltykov method. With an increasing of N the calculated values ξ_{defor} converge to some (real) value for every δ. Computational stability of Monte Carlo algorithm is sufficiently clear from Figure 6.

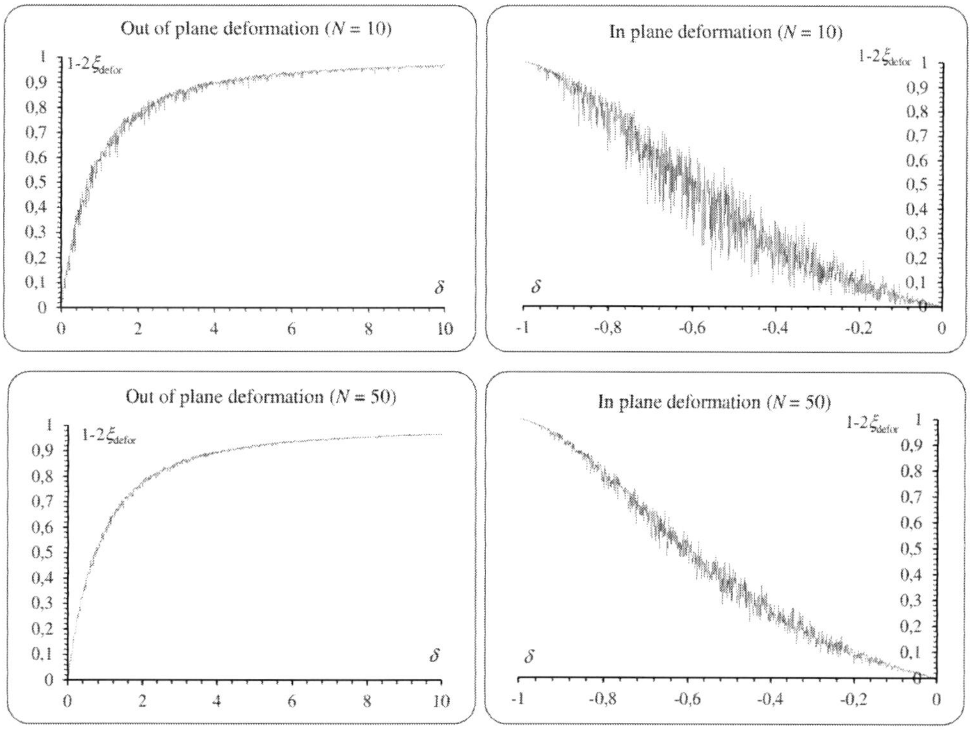

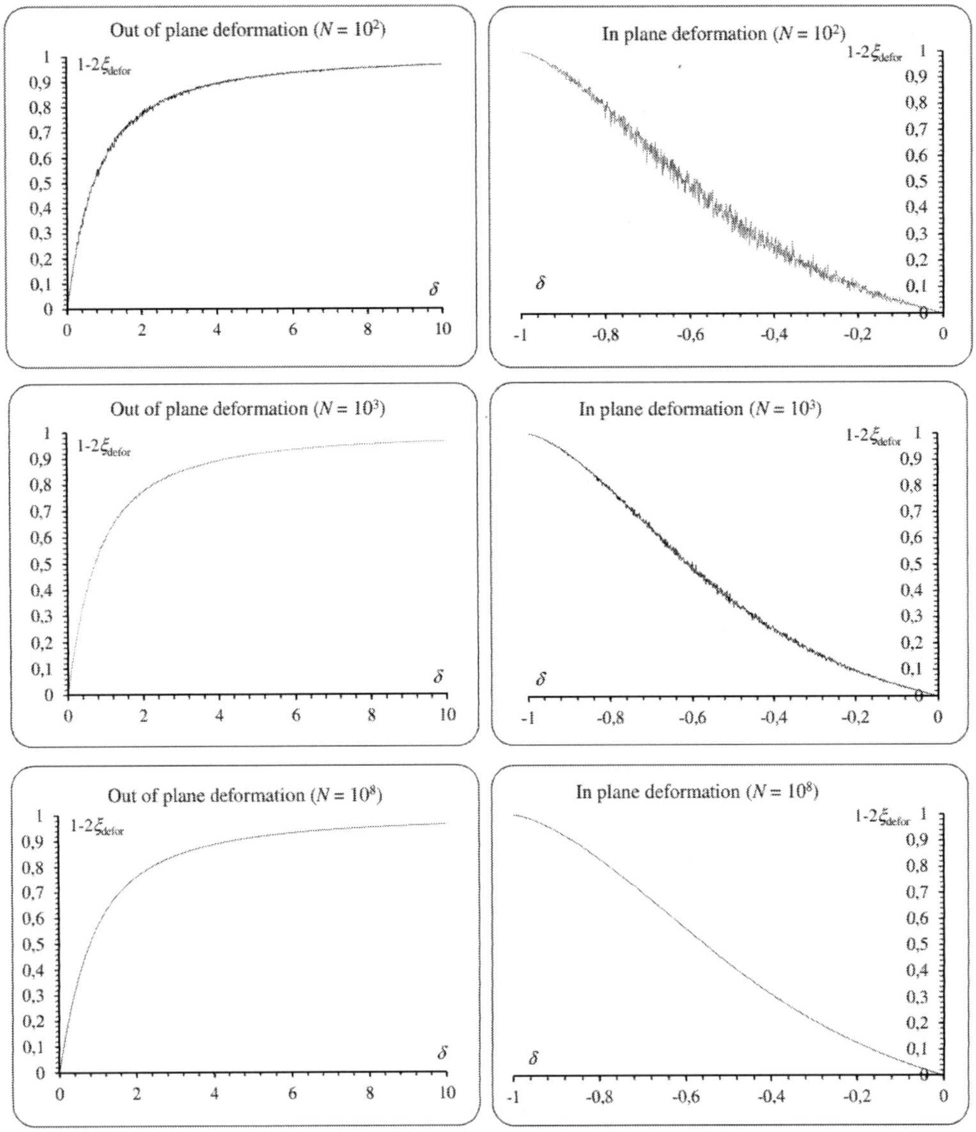

Figure 6. Graphs illustrating Monte Carlo simulation of values ($1\text{-}2\xi_{defor}$) as a function of prolongation ($\delta > 0$) and contraction ($\delta < 0$) for plastically deformed random grained structure.

7. Conclusion

In this work we present brief mathematical treatment of quantities resulting from standard stereological procedures. According to our findings, these quantities can be applicable for the modelling of the local plastic deformation in grained structures. Our approach is based on the assumption that the grain, as it appears on a two-dimensional cross section, will change its shape and spatial orientation during plastic deformation. Obtained results are applied to a homogeneous and isotropic grained structure.

We note that formulas (5.5) and (5.11) can be actually used for the conversion of grained structure orientation represented by quantity ξ into local plastic deformation of the structure characterized by δ. Graphs of the calculated dependence $\xi_{defor}(\delta)$ obtained using these formulas by means of Monte Carlo simulation demonstrate the relevance of stereological methods for the local plastic deformation modelling in grained structures.

Results expressed in (5.5) and (5.11) are not a general solution. It can only be applied to the particular case for which there is no rigid solid rotation of grains in structure and the axes chosen are principal axes of strain (therefore, there are no shear components present). In addition another limitation is imposed, which is that plastic deformation must be isotropic. However it does not seem that a grain with anisotropic behavior would suppose complications in practical applications.
In conclusion, we reported the problem of applicability of stereological methods to determine the local plastic deformation in grained structure. One can easily understand that there is a correlation between the change in the grain orientation and grain deformation. This correlation was mathematically demonstrated in our work using parameter ξ reflecting orthogonal 2D projection of the grained structure onto the oriented plane. Indeed, the expression (3.4) allows one to investigate this correlation.

Acknowledgment
This work was supported by the VEGA Grant No. 1/0122/16 of the Grant Agency of the Slovak Republic Ministry of Education and by the Slovak Research and Development Agency under the contract No. APVV-15-0319.
This work was also funded by the ERDF - Research and Development Operational Programme under the project "University Scientific Park Campus MTF STU - CAMBO" ITMS: 26220220179.

References

[1] Saltykov S A 1970 *Stereometric metallography* Metallurgia, Moscow (in Russian).
[2] Vander Voort G F 2013 *Measuring the Grain Size of Specimens with Non-Equiaxed Grains.* Practical Metallography **50** 4 239-251
[3] Konrad N, Wenxiong H 2004 *Comput Methods Appl Mech Eng* **193** 2719–2743
[4] Ronaldo I B, Atilla A 2004 *Comput Methods Appl Mech Eng* **193** 2667–2698
[5] Wellmann Ch, Wriggers P 2012 *Comput Methods Appl Mech Eng* **205–208** 46–58
[6] Benedetti I, Aliabadi M H 2013 *Comput Methods Appl Mech Eng* **265** 36–62
[7] Martinkovic M, Pokorny P 2015 *Key Engineering Materials* **662** 173 - 176
[8] Martinkovic M, Minarik S 2016 *Short notes on the grains modification by plastic deformation*, in: Hubbard D. (Ed.): Plastic Deformation. NOVA publishers 1-44.
[9] Riosa P R, Glicksmanb M E 2006 *Materials Research* **9** 2 231-236

IOP Conf. Series: Materials Science and Engineering **465** (2019) 012006 doi:10.1088/1757-899X/465/1/012006

Microstructural and mechanical properties of dissimilar Al-Ti joints prepared by GTAW welding-brazing

M Nagy[1], M Behúlová[1] and M R Pérez[2]

[1]Slovak University of Technology in Bratislava, Faculty of Materials Science and Technology in Trnava, Ulica Jána Bottu č. 2781/25, 917 24 Trnava, Slovakia
[2]Universidad Central "Marta Abreu" de Las Villas, Facultad de Ingeniería Mecánica e Industrial, Carretera a Camajuaní Km. 5 y 1/2, Santa Clara, Villa Clara, Cuba

mate.nagy@stuba.sk, maria.behulova@stuba.sk, manuelr@uclv.edu.cu

Abstract. Joining of light metals and alloys including Al-Ti is of great interest of many producers particularly from automotive and aerospace industries. In this paper, conventional gas tungsten arc welding-brazing method was applied to join 2 mm thick plates of wrought AA5083-H111 aluminium alloy with commercially pure titanium Grade 2 using ER4043 filler rod. Butt type of joints with single- and double-square-groove welds were produced using various welding currents. Mechanical properties of prepared welded-brazed joints were evaluated by tensile tests and microhardness measurements. Microstructural analyses were carried out using light microscopy. The weld zones and Al-Ti interfaces were observed by scanning electron microscope. EDX microanalysis was used to identify variations in chemical composition across the welded-brazed joint interface. Finally, the analysis of the influence of welding parameters on the mechanical properties and dimensions of the fusion zone was accomplished.

1. Introduction

Dissimilar joining of light metals and alloys including Al-Ti is of great interest of researchers and producers particularly from automotive and aerospace industries. It results from the combination of increasing demands on production of lightweight but on the other hand high-strength constructions [1-3]. Joining titanium and aluminium as well as their alloys using fusion welding technologies is difficult due to very different material properties and especially because of the formation of brittle intermetallic compounds (IMC) in the interphase layers [4-5]. To produce high-quality weld joints of Al-Ti alloys, the thickness of the brittle intermetallic layer shall be minimized. Many studies have been focused on eliminating the thickness of the IMC layers using different filler metals [1-2, 6], varying the type of joint [4, 7] or by changing the welding method [6-10]. According to the low energy input, technology of gas tungsten arc welding-brazing offers a flexible and relatively inexpensive method for material joining in many industries [1, 3-5, 11].

The main objective of this research was to produce the reliable butt type of joints with the single- and double-square-groove welds from different types of base materials, wrought AA5083-H111 aluminium alloy and commercially pure titanium Grade 2, exploiting the welding-brazing technology of gas tungsten arc welding (GTAW) using ER4043 filler rod in order to study the influence of welding parameters on the microstructure and mechanical properties of welded-brazed joints.

2. Materials and methods

The experimental test samples of dissimilar base materials, the wrought AA5083-H111 aluminium alloy and commercially pure titanium Grade 2 (CP Grade 2) with dimensions of 50 mm × 50 mm × 2 mm were experimentally single- and double-square-groove welded in the flat position using ER4043 filler rod 1.2 mm in diameter. The experimental joints were produced using SELCO Genesis 3200 AC/DC welding machine. The constant welding parameters according to GTAW technology were: the arc voltage $U_v = (12.8 \pm 0.4)$ V, the welding speed $v = 2.5$ mm.s^{-1}, argon shielding gas with the flow rate of 20 L.min^{-1} and the welding mode (2-step welding method with the direct current). As the variable parameters were used the welding current (ranging from 40 A to 60 A) and the number of welds.

The chemical analysis of the base materials was carried out using JEOL 7600 scanning electron microscope equipped with EDX analyzer and compared with the data from data sheets [8]. The chemical composition of welded material is given in Table 1 together with the chemical composition of the filler material specified according to the material data sheet [11-12].

Table 1. Chemical composition of experimental materials in [wt.%].

Element	Mg	Si	Cr	Cu	Mn	Fe	O	N	H	Ti	Al
CP Grade 2	-	-	-	-	-	-	0.25	0.03	0.015	Bal.	-
AA5083	4.7	0.4	0.13	0.19	0.26	0.31	-	-	-	-	Bal.
ER4043	0.05	4.5 - 6	-	0.3	0.05	0.8	-	-	-	0.2	Bal.

The measured mechanical properties of the base materials and reference mechanical properties of the filler material are summarized in Table 2 [12].

Table 2. Mechanical properties of experimental materials.

	Tensile strength [MPa]	Young modulus [MPa]	Elongation [%]	Yield strength [MPa]	Microhardness [HV0.1]
CP Grade 2	441 ± 3.5	105	20	330	178 ± 3
AA5083	267 ± 2	73.35	29.7	180 ± 2	75 ± 2
ER4043	190	74	8	164	-

The experimental weld joints prepared by GTAW technology were investigated using mechanical testing, macrostructural and microstructural observations and the chemical analysis in the region of the Ti and weld metal (WM) interface. The test specimens were prepared by milling machine from the center of experimentally welded-brazed joints.

The destructive transverse tensile tests of experimental welds were performed according to standards STN EN ISO 6892-1 [13] and STN EN ISO 4136 [14] applying the LabTest 5.250 SP1-VM testing machine using the testing speed 1 mm.s^{-1}. The microhardness measurements were performed according to STN EN ISO 6507-1 [15] and STN EN 9015-2 [16] standards applying the Buehler IndentaMet 1100 microhardness testing machine. The testing load was HV 0.1 acting during 10 s. The distance between indents was 250 µm.

For macroscopic and microscopic examination, the metallographically prepared joint cross sections were etched by Keller´s etchant with the composition in volume and in order of mixing: 950 ml water (H_2O) 25 ml nitric acid (HNO_3), 15 ml hydrochloric (HCl), 10 ml hydrofluoric acid (HF) etching for 3 to 5 s [17]. The macroscopic analysis of samples was performed using ZEISS Stemi 2000-C equipment with the AxioCam ERc5s camera. The sample microstructures were observed by NEOPHOT 32 light microscope and the JEOL 7600 scanning electron microscope equipped with EDX analyzer. The microstructures of base materials are illustrated in Figure 1. The CP Grade 2 microstructure consists of fine equiaxial α grains with the average size of 20 ± 5 µm. The microstructure AA5083-H111 alloy is formed by αAl solid solution with dispersed Mg particles.

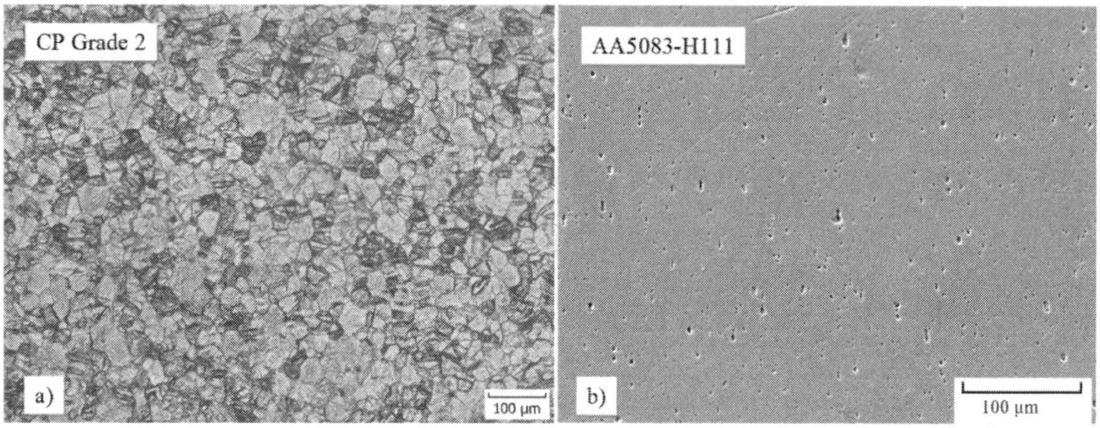

Figure 1. Microstructure of the base materials a) titanium CP Grade 2 and b) AA5083-H111 alloy.

3. Results and discussion

Butt type of joints with single- and double-square-groove welds were experimentally GTAW welded-brazed using ER4043 filler rod and welding parameters given in Table 3. The appearance of produced experimental joints is illustrated in this table as well. During the examination of the joint quality, the test samples from welded-brazed joints were subjected to mechanical strength tests and microhardness measurements. The macroscopic investigation of the weld quality was completed with the measurements of the weld metal widths (W_{WM}) and weld metal areas (S_{WM}). The microstructural characterization and EDX chemical analysis of the WM and Ti-WM interface was performed applying the experimental sample number 5.

Table 3. Experimental parameters and appearance of prepared welded-brazed joints.

	Weld	I_A [A]	U_v [V]	v [mm.s^{-1}]	Top of the weld	Weld root
1	Single	40	12.7	2.5	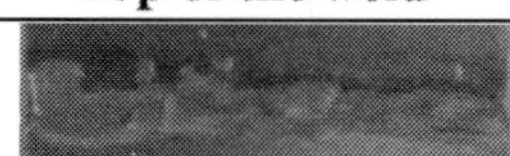	
2	Single	50	12.6	2.5		
3	Single	60	12.9	2.5		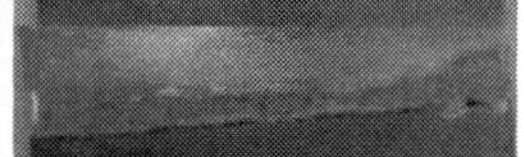
4	Double	40	13	2.5	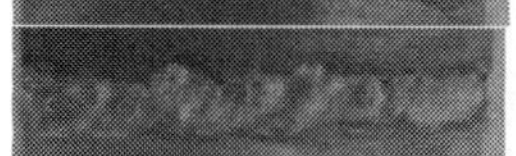	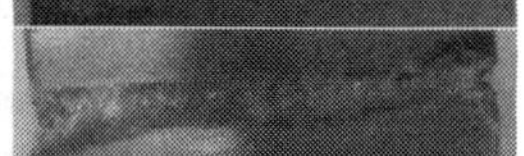
5	Double	50	12.7	2.5		
6	Double	60	12.8	2.5	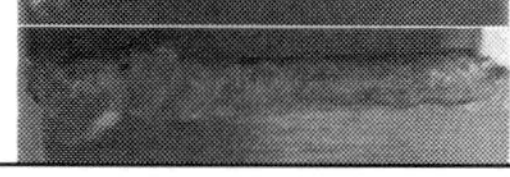	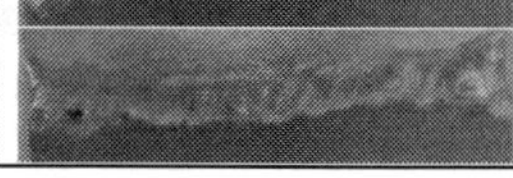

3.1. *Mechanical testing*

Results of the tensile strength tests of produced dissimilar joints are shown in Figure 2. All samples tested by tensile strength tests were broken in the WM. The ultimate tensile strength (R_m) of samples with the double-square-groove welds are higher than the values measured for single-square-groove welds. The highest value of the ultimate tensile strength was reached for the sample 5 with a double-square-groove weld prepared using the welding current of 50 A. The achieved value of 153 MPa represents 57 % of the ultimate tensile strength of the AA5083-H111 base material.

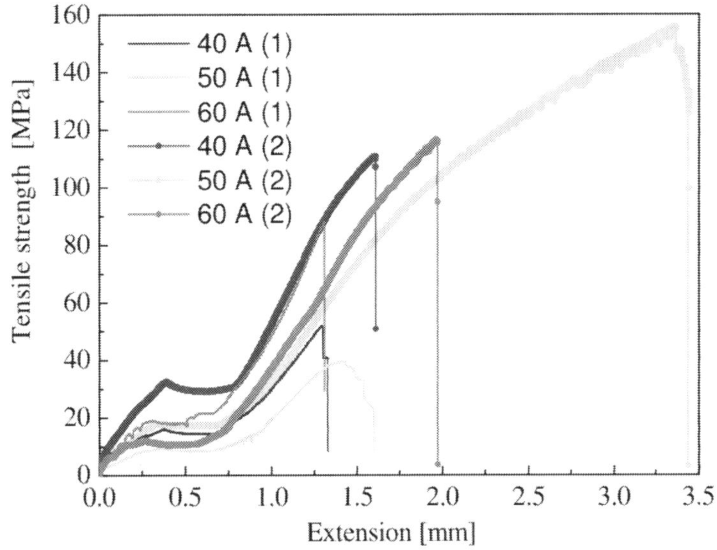

Figure 2. Tensile strength tests of experimental joints.

Results of microhardness examination of experimental samples across the weld centreline are plotted in Figure 3. On the titanium side, for the samples with single-square-groove welds the higher values of microhardness were measured in comparison with microhardness of the CP Grade 2 base material (Table 2). It can be supposed that the increase in microhardness results mainly from the martensitic and polymorphic transformations typical for Ti alloys [2]. The decrease in microhardness of titanium CP Grade 2 base metal in samples with double-square-groove welds prepared with higher currents is probably related to the tempering effects during formation of the second weld. The microhardness in the WM zone is higher than the microhardness AA5083-H111 base material due to the intermixing of molten aluminium alloy and the filler material. The lowest microhardness of WM was measured for the sample 2 with the single-square-groove weld produced using the welding current of 50 A. The microhardness of AA5083-H111 base material in a heat affected zone (HAZ) slightly decreases as a results of the grain size increase [4-6].

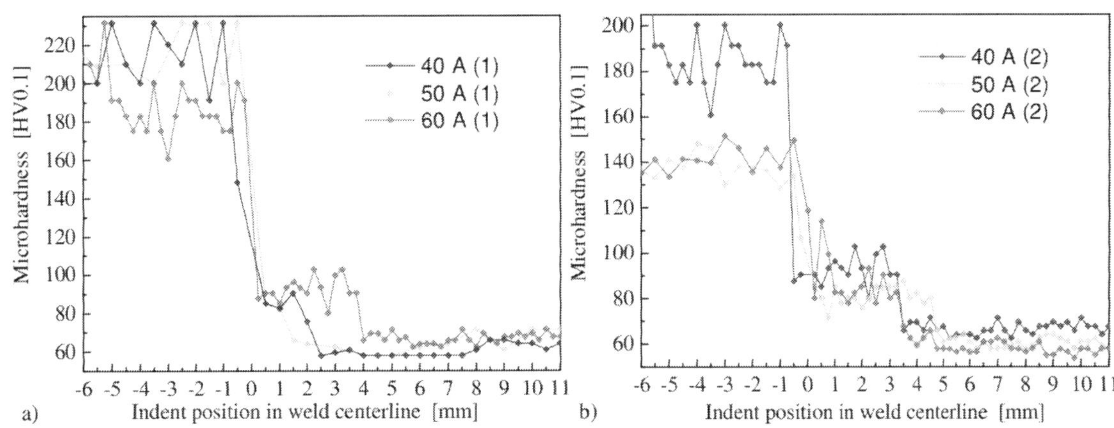

Figure 3. Microhardness measurements across the weld centerlines of a) single-square-groove welds and b) double-square-groove welds prepared using different welding currents.

3.2. Macroscopic analysis

Macrostructures of experimental joints are shown in Figure 4. The WM is asymmetric with a prevailed fusion extent on the Al side resulting from the essentially lower liquidus temperature of Al alloy compared to the melting temperature of CP Grade 2. Some geometrical imperfections such as incomplete root penetration (Figure 4a, 4d) and linear misalignment between plates (Figure 4b, 4d) were found.

On the cross-sections from the experimentally produced welded-brazed joints, several zones characteristic for welding-brazing of dissimilar materials can be identified, i. e. a brazed joint on the titanium side, weld metal and the typical weld joint at the aluminium alloy side. The not melted or partially melted zone at the Ti and WM interface is characteristic for dissimilar joints prepared by brazing. The applied heat input during GTAW process was not sufficient to melt the titanium sheet along the whole thickness. The molten filler material only wetted the titanium sheet resulting in the brazed joint formation on the titanium side. On the AA5083 aluminium alloy side, melting of the base metal and also filler material occurred, producing a weld joint with the as-cast microstructure.

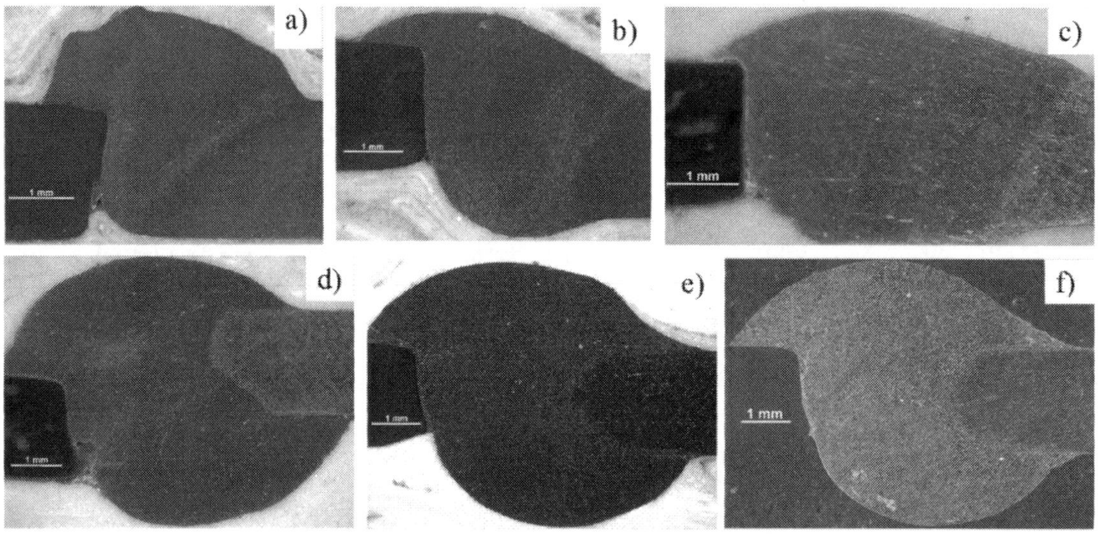

Figure 4. Macrostructures in cross-sections of experimental joints.

Using macrostructures in Figure 4 and results of microhardness testing (Figure 3), the widths of WM in the middle of the welded-brazed joints and the areas of WM were measured. The results are summarized in Table 4 together with the welding-brazing process parameters, measured ultimate tensile strength R_m and the heat input P and linear heat input Q_l computed according to relationships

$$P = U I \eta \qquad [W] \qquad (1)$$

$$Q_l = \frac{P}{v} \qquad [J.m^{-1}] \qquad (2)$$

in which the efficiency η was supposed to be 75 % [11]. With increasing heat input and/or linear heat input the width of WM W_{WM} and also the WM area S_{WM} enlarge owing to the increase in volume of molten filler material.

3.3. Microscopic analysis

The microscopic analysis of the Al HAZ-WM and Ti-WM interfaces was carried out for the experimental sample no. 5 exhibited the highest tensile strength. Figure 5a illustrates the microstructure in the area of the interface between WM and the HAZ at the AA5083-H111 side.

Table 4. Experimental welding parameters, calculated and measured results.

	Weld	I_A [A]	U_v [V]	P [W]	Q_l [J.mm^{-1}]	R_m [MPa]	W_{WM} [mm]	S_{WM} [mm^2]
1	Single	40	12.7	381	152.4	52	2.5	6.99
2	Single	50	12.6	472.5	189.0	40	3	9.30
3	Single	60	12.9	580.5	232.2	88	3.75	11.95
4	Double	40	13	780	312.0	111	4.25	20
5	Double	50	12.7	952.5	381.0	156	4.5	21.34
6	Double	60	12.8	1152.0	460.0	117	4.75	22.96

In the fusion zone, the mixing of the molten Al base alloy and filler material occurred leading to the enrichment of weld metal by Si. The grain refinement in the WM and the change in grain morphology at the interface Al HAZ-WM can be also observed in Figure 5a. Figure 5b demonstrates the Ti-WM interface formed by quasi-continual IMC layer in the upper part of the welded-brazed joint where the Ti base metal was partially melted. Columnar grains connected to this layer are oriented towards the weld centre creating the dispersed non-continuous layer. In the bottom part of the welded-brazed joint, it is possible to observe continuous oxide layer with micro-pores along the Ti-WM interface. A large pore developed by Ar capture in WM is visible in Figure 5b as well.

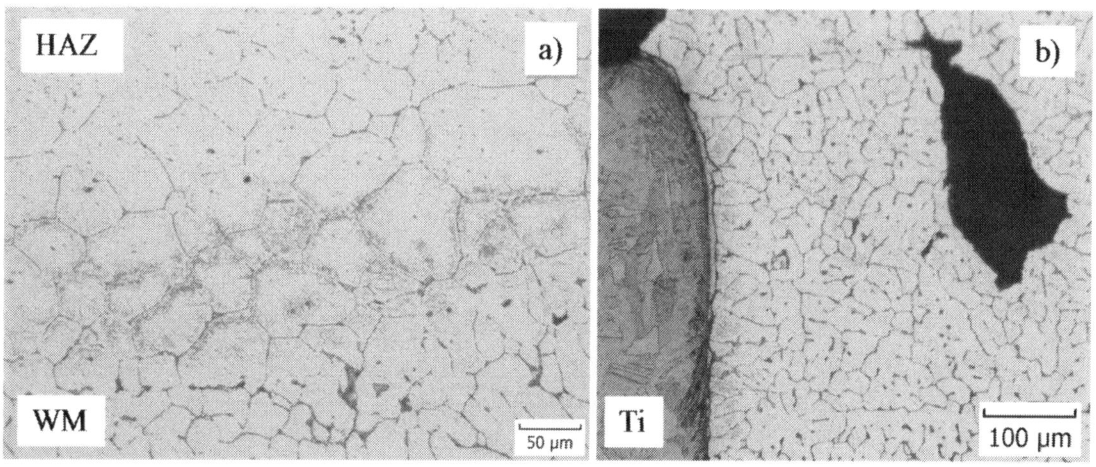

Figure 5. Microstructure of a welded-brazed joint in the area of a) HAZ-WM interface and b) Ti-WM interface.

3.4. EDX analysis

The EDX analysis in the area of Ti-WM interface was carried out on sample 5. The distribution maps of chemical elements Al, Ti, Si, Mg, Fe and O in the Ti-WM region are shown in Figure 6. The results of EDX line analysis are plotted in Figure 7. The line scan analysis from the WM toward the Ti side shows that the Al content falls and the Ti content rises relatively sharply to a constant level. The detail of the line scan from WM (upper diagram) suggests that WM is formed mainly by Al-Si eutectics. This assumption corresponds to the previously obtained results [2, 4-6]. At the WM-Ti interface the concentration of Mg drops down. The slightly increased concentration of the oxygen indicate the presence of an oxides layer at the WM-Ti interface.

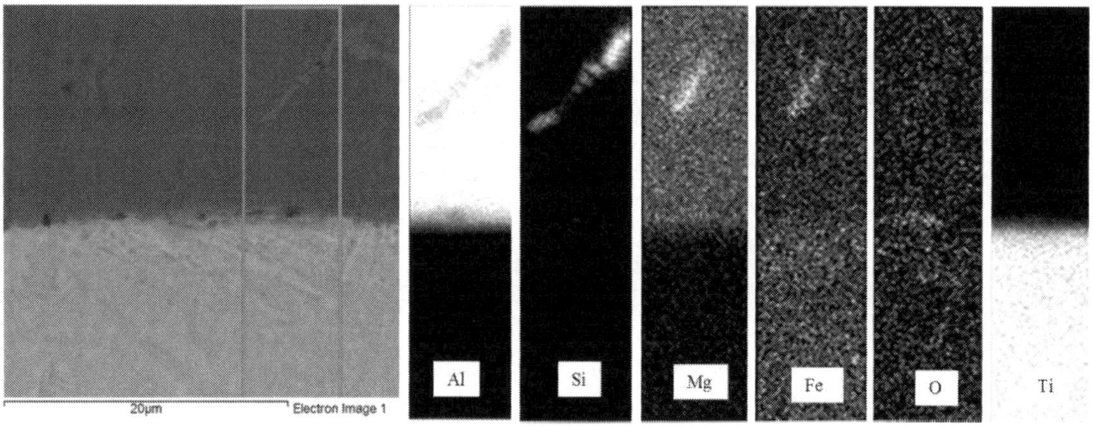

Figure 6. Distribution maps of chemical elements in the area of the Ti-WM interface.

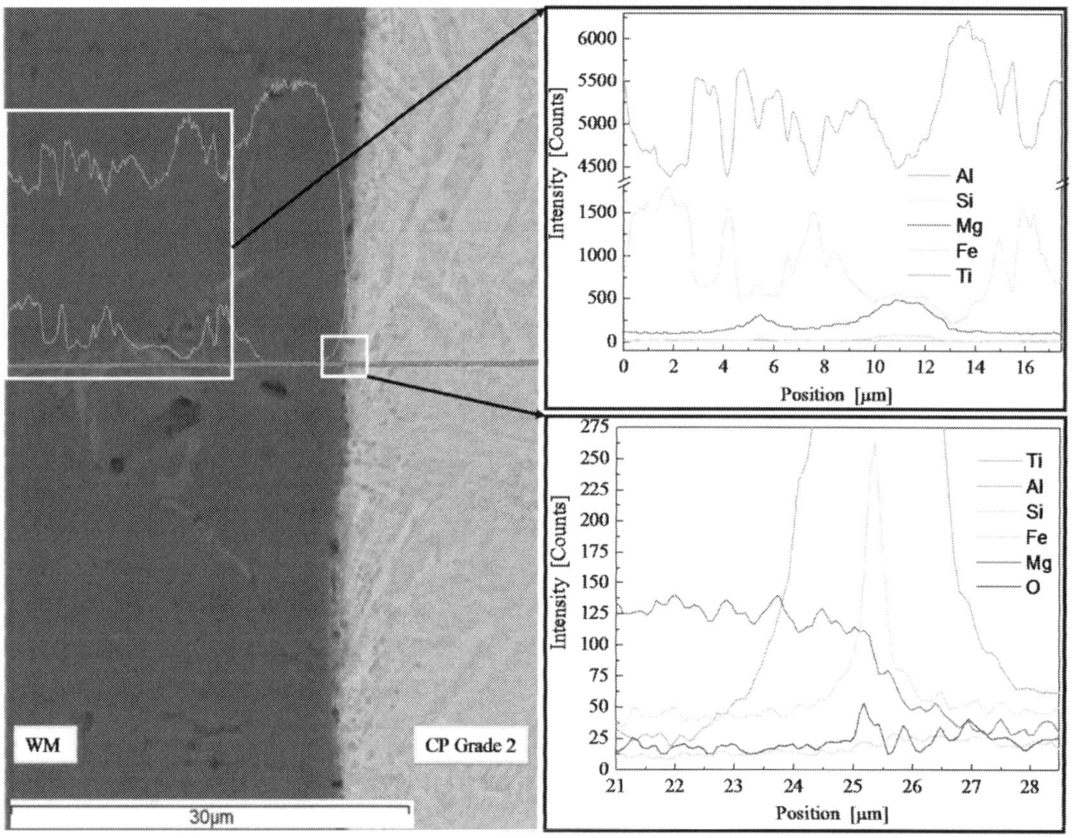

Figure 7. Concentration profiles of chemical elements across Ti-WM interface.

The selected spectra in the points near the Ti-WM interface according to Figure 8 were evaluated to identify the chemical composition and possible phases. The spectra from 1 to 4 and 8 were measured in the WM while the spectra from 5 to 7 at the WM-Ti interface and in Ti side, respectively. According to phase diagrams and published results, the possible phases indicated in Table 5 were proposed.

Table 5. EDX analysis [at.%] in the points according to Figure 8.

Spectrum	Al	Ti	Si	Mg	Fe	Cu	Mn	Ar	Possible phases	References
1	97.8	0.11	0.97	0.21	0	0	0	0.91	Al matrix	[19]
2	63.19	0.08	32.76	3.27	0.2	0.1	0	0.39	Al_5Cu, Mg_2Si	[20-22]
3	86.3	0	12.63	0.56	0	0	0	0.51	AlMgSi	
4	69.46	0	19.71	8.1	2.18	0	0.17	0.39	Mg_2Si, Mg_2Al_3, $Al_{12}Mn$, AlMgFeSi	[18, 20-22]
5	6.14	92.83	0.77	0	0.26	0	0	0	Ti_3Al, Ti_3Si	[20, 5-6]
6	0.47	99.53	0	0	0	0	0	0	Ti_3Al, αTi	[8, 20]
7	42.08	55.87	1.88	0	0	0	0	0.17	TiAl, Ti_5Si_3	[8, 5-6]
8	76.69	0.1	22.3	0.34	0	0	0	0.57	AlMgSi	[19-21]

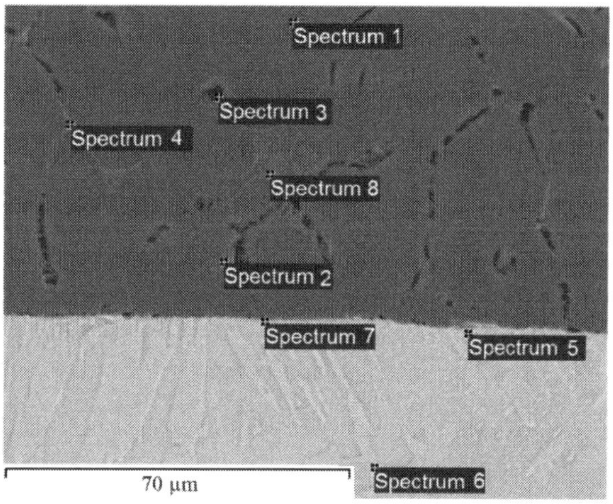

Figure 8. Positions of EDX spectra.

4. Conclusions

The sheets of AA5083-H111 aluminium alloy and titanium CP Grade 2 were experimentally butt-welded-brazed using conventional GTAW welding-brazing method with separately added ER4043 filler rod. Using the welding current ranging from 40 A to 60 A, single- and double-square-groove welds were prepared. The quality evaluation of welded-brazed joints was carried out applying mechanical strength tests, microhardness examination and microstructure characterization including EDX analysis.

All experimental samples from the welded-brazed joints tested under uniaxial tensile loading were broken in the weld metal. The maximum ultimate tensile strength at the level of 153 MPa was achieved for the sample 5 produced with the welding current of 50 A and double-square-groove weld. This value represents the 57 % of AA5083-H111 base material ultimate tensile strength. The lowest microhardness of weld metal was measured for the sample 2 with the single-square-groove weld prepared using the welding current of 50 A.

Macroscopic examination of welded-brazed joint cross-sections revealed the asymmetric shape of the fusion zones with a larger extent of molten material on the Al alloy side. Some geometric imperfections such as incomplete root penetration and linear misalignment between welded-brazed sheets were found. The interface between WM and the CP Grade 2 base material was formed mainly by brazing with solid-liquid interface while the joint on the Al alloy side exhibited the typical features of fusion welding. With increasing welding current or heat input the width of WM as well as the area of WM increased.

The microscopic analysis of the Al HAZ-WM and Ti-WM interfaces was carried out for the experimental sample no. 5 exhibited the highest ultimate tensile strength. In the area of Ti-WM interface, the EDX analysis was performed as well.

Acknowledgements
The research has been supported by the Scientific Grant Agency of the Slovak Republic within the VEGA project No. 1/1010/16, KEGA project No. 029STU-4/2018 and the project ITMS 26220120048 under the Research & Development Operational Programe funded by the ERDF.

References

[1] Shixiong L, Qinglong C, Yongxian H and Xiaojun J 2013 *Mat Sci Eng A-Struct* **568** 150-154
[2] Ma Z, Wang Ch, Yu H, Yan J and Shen H 2013 *Mater Design* **45** 72-79
[3] Kun L, Yajiang L, Shouzheng W and Juan W 2014 *Mater Manuf Process* **29** 969–974
[4] Haodong W, Xinjian Y, Ting L, Kanglong W, Yongqiang S and Chuan X 2018 *J Mater Process Tech* **251** 26-36
[5] Liu K, Li Y, Wang J, Wei S and Zhou Y 2016 *Kovove mater* **54** 97-105
[6] Chen S, Li L, Chen Y and Liu D 2010 *Trans Nonferrous Met Soc China* **20** 1 64-70
[7] Baqer Y M, Ramesh S and Yusof F 2018 *Int J Adv Manuf Technol* **95** 4353
[8] Sahul M, Sahul M, Vyskoc M, Caplovic D and Pasak M 2017 *J Mater Eng Perform* **26** 3 1346-1357
[9] Casalino G, Mortello M and Peyre P 2015 *J Mater Process Tech* **223** 139-149
[10] Sun Q J, Li J Z, Liu Y B, Li B P, Xu P W and Feng J C 2017 *Mater Design* **116** 316-324
[11] ASM International Handbook Committee 1993 ASM Handbook *Welding, Brazing, and Soldering* ASM International USA vol **6** p 2873
[12] ASM International Handbook Committee 1990 ASM Handbook *Properties and Selection: Nonferrous Alloys and Special-Purpose Materials* ASM International USA vol **2** p 3470
[13] Destructive tests on welds in metallic matrials – Transverse tensile tests (ISO 4136:2012)
[14] Metallic materials – Method of test at room temperature (ISO 6892-1:2016)
[15] Metallic matrials. Vickers hardness test. Part 1: Test method (ISO 6507-1:2005)
[16] Destructive tests on welds in metallic matrials – Hardness testing – Part 2: Microhardness testing of welded joints (ISO 9015-2:2016)
[17] Destructive tests on welds in metallic matrials – Etchants for macroscopic and microscopic examination (ISO/TR 16060:2003)
[18] Cao R, Sun J H and Chen J H 2011 *Sci Technol Weld Joi* 5 **18** 425-433
[19] Sambasiva R A, Madhusudhan R G and Satya P K 2011 *Mater Sci Tech-Lond* **27** 1 65-70
[20] Batalu D, Coşmeleaţă G and Aloman A 2006 *U P B Sci Bull* B **68** 4 77-90
[21] Biao H, Song Q, Yong D, Zhiyong L and Qingping W 2015 *J Phase Equilib Diff* **36** 4 333-349
[22] ASM International Handbook Committee 1991 ASM Handbook *Alloy Phase Diagrams* ASM International USA vol **3** p 1741

IOP Conf. Series: Materials Science and Engineering **465** (2019) 012007 doi:10.1088/1757-899X/465/1/012007

Evaluation of Material Deformation During Process of Precise Carbon Steel Tube Cold Draw Forming

M Necpal and M Martinkovič

Slovak University of Technology in Bratislava, Faculty of Materials Science and Technology in Trnava, Ulica Jána Bottu č. 2781/25, 917 24 Trnava, Slovakia

martin.necpal@stuba.sk, maros.martinkovic@stuba.sk

Abstract. A cold forming technology (cold drawing) produces precise seamless tubes, which can be found in various industries and applications. A high shape accuracy is achieved while the material is reinforced due to plastic deformation during cold forming. The reinforcement of the material in the forming process largely depends on the degree of deformation and plastic properties of the material. The purpose of this study is to compare results of stereological methods used in metallography with data from FEM simulation obtained using software DEFORM and Lagrangian method. Values of strain obtained experimentally and by FEM simulations are supplemented by results of Vickers hardness measurements and mechanical properties obtained by tensile testing after each cold draw operation.

1. Introduction

Cold forming is widely applied process in industrial production. High-quality tubes under acceptable economic conditions are manufactured by draw process. To achieve the final diameter and wall thickness of tubes this process is reduced successively to several cold drawing steps. Draw process can be done by either drawing the tube through a die and adding a plug, which results in better-defined surface quality. As a feedstock for cold drawing is usually used the hot rolled tube. During subsequent drawing steps, the tube undergoes plastic deformation, generating residual stresses and strain. Palkowski at al. [1] investigated residual stress state of drawn tubes by numerical simulation and neutron diffraction analysis. The material of drawing tube was a cooper. Neves at al. [2] were looking for the best geometry of die and plug to reduce the drawing force just using the FEM simulation. The simulation result has been compared with experiment. Zhang at al. [3] investigated surface roughening analysis of cold drawn tube. The micro-scale finite element model with initial surface roughness was used for simulation. Toribio at al. [4] analyzed the influence of hydrogen embrittlement on parameters of the wire drawing process using various FEM numerical simulations.

This paper is focused on the comparison of the result of the strain values obtained by experiment and FEM simulation using software DEFORM 2D and Lagrangian formulation of simulation problem.

2. Experiment

Steel grade E235 and E355 (ferritic-pearlitic carbon steels with the chemical composition given in Table 1) were investigated experimentally; as a feedstock for cold drawing a hot rolled tube with dimensions of Ø70 × 6.3 mm was selected.

Mechanical properties of E235 steel can be summarized as follows: yield stress $R_e = (226 \div 245)$ MPa, tensile strength $R_m = (343 \div 441)$ MPa and ductility $A_5 = 24$ %. Mechanical properties of E355 steel are: yield stress $R_e = (335 \div 355)$ MPa, tensile strength $R_m = (490 \div 630)$ MPa and ductility $A_5 = 24$ % [6].

Table 1. Chemical composition of E235 and E355 steel grade in wt. % [5].

	C	Mn	Ni	Cu	Mo	Sn	P	S	Si	Ti	Cr
E235	0.090	0.420	0.060	0.140	0.020	0.012	0.013	0.011	0.200	0.002	0.060
E355	0.180	1.220	0.110	0.170	0.030	0.017	0.025	0.012	0.270	0.002	0.100

Chemical treatment of the tube (pickling, phosphating, lubricating) was made before cold drawing process. Geometrical and technological parameters for experimental tube drawing are shown in Table 2. Calculation of tube cross-section (area) reduction come given by

$$Reduction = \frac{S_R}{S_0} \cdot 100 \, [\%], \tag{1}$$

where the ($S_R = S_0 - S \, [mm^2]$) ,($S_0 \, [mm^2]$) is the tube area before drawing and ($S \, [mm^2]$) is tube area after drawing.

In the 1st pass the area reduction was 40.52 %, and in the 2nd pass, the area reduction was 27.34 %. The final reduction of the tube cross section area for two-pass technology (from Ø70 × 6. 3 mm to Ø50 × 3.75 mm) was 57.78 %; the value was divided rather uniformly among all passes (see Table 2).

Table 2. Geometrical and technological parameters for experimental tube drawing.

	Feedstock dimensions [mm]			**Tube dimensions [mm]**			**Pointing [mm]**		**Reduction [%]**		
Pass No.	O.D.	W.T.	L	O.D.	W.T.	L	L (tip)	Ø (tip)	Total	W.T.	O.D.
1	70	6.30	4300	55	4.75	6725	300	40	40.52	25	21
2	55	4.75	6725	50	3.75	9255	300	40	27.34	21	9

O.D.-outer diameter, W.T.-wall thickness, L-length

The local plastic deformation of the tube was analyzed. During subsequent drawing steps, the tube undergoes plastic deformation. Final geometry and properties of deformed tubes depend on strain value. It is possible to investigate the change of material structure and predict plastic deformation.

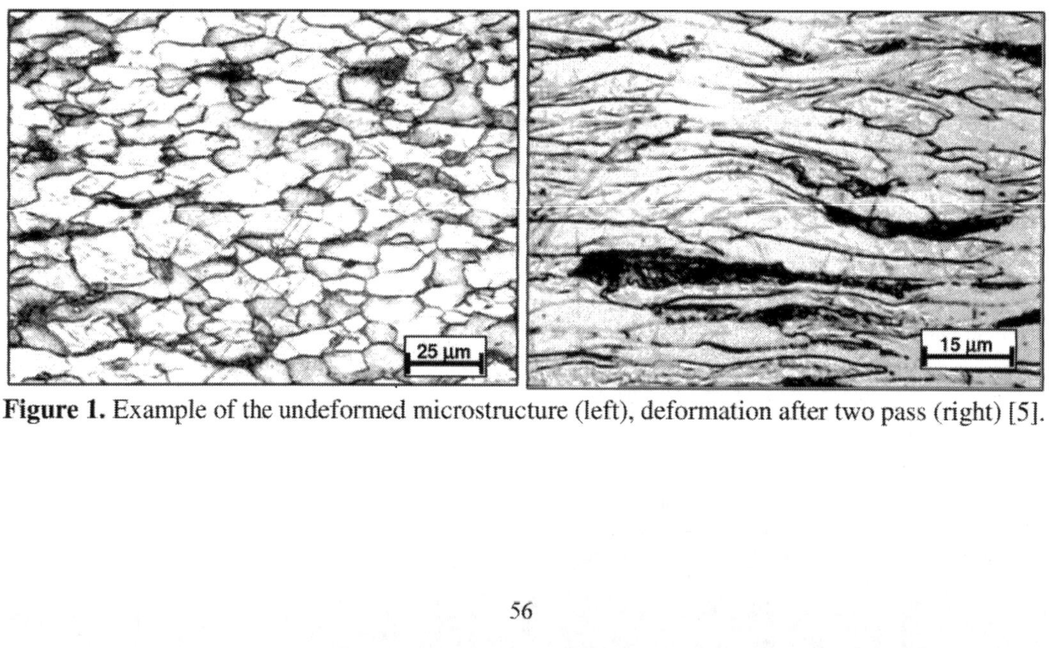

Figure 1. Example of the undeformed microstructure (left), deformation after two pass (right) [5].

The microstructure after both passes was investigated in orthogonal tangential and longitudinal planes of the deformed tube by light microscope (500× magnification) on metallographic cut (Figure 1). Three tangential metallographic cuts were prepared: 1. on the inner surface of the pipe wall "I", 2. on the outer surface of the pipe wall "O" and 3. in the middle of the pipe wall "M" (Figure 2). The structure anisotropy increased due to plastic deformation – grain boundaries orientation in deformed area of the piece was evaluated. The local strain in analyzed area was obtained by stereological measurement of the degree of grain boundaries orientation [7]. The anisotropic microstructure was decomposed into isotropic and planar oriented components using quantitative metallography [8] (Saltykov stereology methods with oriented test lines). Measured and calculated parameters of local plastic deformation in the deformed zone are summarized in Table 4.

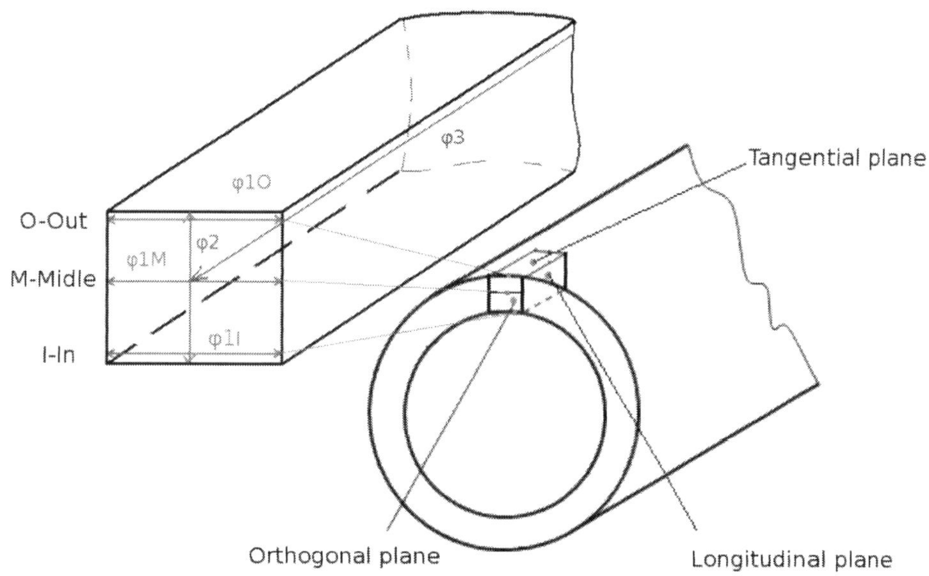

Figure 2. Metallographic cuts and planes on the part of the deformed tube.

3. Numerical simulation

DEFORM 2D and the Lagrangian calculation has been set for numerical simulation of cold drawing technology. The initial axisymmetric geometric configuration has been set. The data interpolation functionality was used to set the second pass calculation. The final data from the first pass was transformed and set the second pass. This data transformation must be realized because drawing technology was performed without interoperative annealing.

3.1. Workpiece and tools of material modelling

As the temperature stays lower in cold drawing, the material properties do not vary and can therefore be supposed to by independent of the temperature variation. As the material behavior is considered temperature independent, the power law constitutive equation of material plastic properties can be used

$$\sigma = c\varepsilon^n\dot{\varepsilon}^m + y \qquad (2)$$

where ε represents the equivalent effective plastic strain, $\dot{\varepsilon}$ represent the effective plastic strain rates, and c, n, m, y material parameters are determined by means of tensile tests [5], they are listed in Table 3. The tube material was considered as plastic, the hardening is assumed as isotropic, and yield function type is set as Von Mises. Elastic material properties can be neglected, the workpiece is set as a plastic.

Table 3. Materials parameters of the stainless steels for Eq. 2.

	Parameter	E235 material	E355 material
c	Material constant (-)	178	404
n	Strain exponent (-)	0.61	0.63
m	Strain rate exponent (-)	0.02	0.02
y	Initial yield value (MPa)	235.5	345

3.2. Tools geometry, properties and simulation condition

The drawing velocity was set 9 m.min^{-1} for all passes in experimental drawing and simulation boundary conditions. The forming tools, (mandrel and die) was considered as rigid bodies. Tool dimensions are given in Table 2. The friction model between the tools and the material was chosen to be a shear-type with the value of 0.08. The geometry of the tube has meshed. The eight elements are across the wall of the tube is in the first pass the six are in second pass Figure 4. This mesh of workpiece is sufficient for accuracy and does not need to be re-mesh procedure during the calculation. The mesh of tools did not need to be generated for simulation

4. Result and discussion

Comparison of strain values measured by the stereological method with values obtained from the simulation was possible. The strain in all directions was analysed according to Figure 2. The φ1 strains were examined in three sections of the cross-section of the tube wall (Figure 3). Strain φ1I is near the inner surface, strain φ1O is near the outer surface, and strain φ1M is in the middle of the tube wall. As shown in Figure 3, after the 1st pass the difference of strain between the E235 and E355 is minimal. Smaller negative strain was identified on the inside of the tube "A" then on the outside of the tube "B". The same result was obtained after 2nd pass.

Calculated values of radial strain φ2 for all passes and materials can be seen in Table 4. Due to high reduction 40.52 % after the first pass, there is an obvious positive φ2 strain in deformed area. After the second pass, where the reduction area is 27.34 %, there is an only negative φ2 strain. Calculated values of φ3 strain for all passes and materials are shown in Figure 4.

The initial strain of undeformed tube is 0. After the first pass φ3 strain was approximately 0.54 for E235 and also E355 material (Figure 4 left). Data interpolation method was used for initialization of the second pass. Initial strain for the second pass was between 0.550 and 0.625 (Figure 4 right). Mechanical properties from tensile testing and Vickers hardness for both materials are shown in Figure 5.

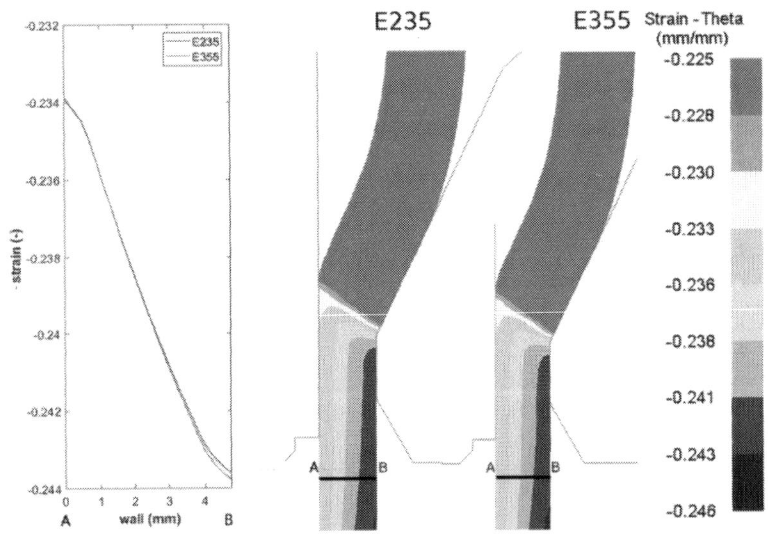

Figure 3. Strain φ1 across the wall of the tube after the 1st pass.

Table 4. Stereology methods and simulation strain values result.

		Values obtained by stereological measurement [5]					Values obtained by numerical simulation				
		φ1O	φ1M	φ1I	φ2	φ3	φ1O	φ1M	φ1I	φ2	φ3
E235	Pass 1	−0.241	−0.237	−0.215	−0.282	0.520	−0.243	−0.239	−0.232	−0.298	0.537
	Pass 2	−0.336	−0.320	−0.301	−0.519	0.839	−0.339	−0.324	−0.305	−0.551	0.874
E355	Pass 1	−0.241	−0.237	−0.232	−0.282	0.520	−0.243	−0.239	−0.233	−0.297	0.536
	Pass 2	−0.336	−0.320	−0.300	−0.519	0.839	−0.339	−0.324	−0.305	−0.55	0.874

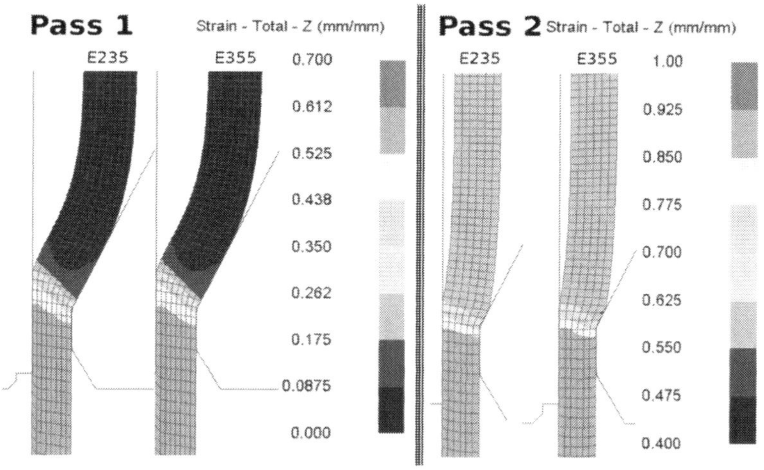

Figure 4. Pass 1 and pass 2 strain φ3 for E235 and E355 steels.

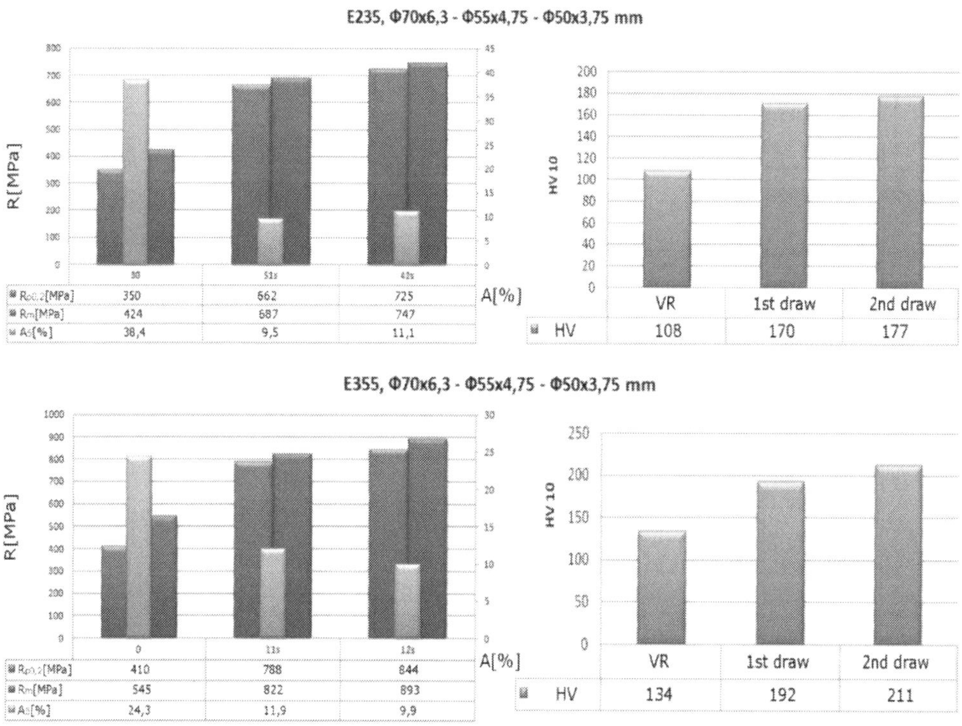

Figure 5. Mechanical properties: tensile testing (left) and Vickers hardness HV (right).

The hardness measured after the second pass was 177.2 HV for E235 material and 211.4 HV for E355 material. Tensile test after the second pass was higher for E355 material (893 MPa) than for E235 steel (747 MPa).

5. Conclusion

The key topic of this paper is to compare parameters of plastic deformation obtained experimentally by measurement of microstructural characteristics of oriented grains with data obtained from FEM simulation using axial-symmetric Lagrangian formulation. Excellent agreement of final strain values from FEM simulation and stereological measurement was achieved. FEM simulation allows analyzing many other parameters of the forming process. Deformation tensor, stress tensor, tool load, tool wear, etc. It is necessary to define correctly the simulation boundary conditions, geometry and material properties for Compliant and Correct Results. Defining tools as rigid and defining a forming material as plastic using Power Law definition appears to be sufficient for simulation of the cold drawing process.

Acknowledgment

This work was supported by the Slovak Research and Development Agency under the contract No. APVV-15-0319 and by the VEGA Grant No. 1/0122/16 of the Grant Agency of the Slovak Republic Ministry of Education.

References

[1] Palkowski H, Brück S, Pirling T and Carradò A 2013 *Materials (Basel)* **6** 11 5118–5130
[2] Neves F O, Button S T, Caminaga C and Gentile F C 2005 *J. Brazilian Soc Mech Sci Eng* **27** 4 426–431
[3] Zhang L, Xu W, Long J and Lei Z 2015 *J Mater Process Technol* **224** 189–199
[4] Toribio J, Lorenzo M, Vergara D and Kharin V 2014 *Eng Fail Anal* **36** 215–225
[5] Martin R 2010 *Research on technological parameters affecting the production and properties* Slovak University of Technology, Bratislava
[6] Fürbacher I, Macek K and Steidl J 1999 *Lexikon technických materiálů* Volume 1. Praha: Verlag Dashöfer
[7] Saltykov S A 1970 *Metallurgia* **3** 446
[8] Martinkovič M and Minarik S 2016 Short notes on the grains modification by plastic deformation *Hubbard D Plast Deform Nov Publ* 1–44

IOP Conf. Series: Materials Science and Engineering **465** (2019) 012008 doi:10.1088/1757-899X/465/1/012008

Enhanced room temperature exciton photoluminescence of plasma hydrogenated ZnO nanocolumns

Z Remes[1], Y Y Chang[1,2], J Stuchlík[1], N Neykova[1], J Soucek[2], Hua Shu Hsu[3]

[1]Institute of Physics of the Czech Academy of Sciences, Cukrovarnicka 10, 16200 Praha 6, Czech Republic
[2]Faculty of Biomedical Engineering CTU, Nam. Sitna 3105, 27201 Kladno, Czech Republic
[3]National Pingtung University, Minsheng Rd. 4-18, 900 Pingtung City,Taiwan

remes@fzu.cz

Abstract. Zinc oxide (ZnO) nanocolumns have interesting material properties such as the direct band gap, tunable electrical conductivity, high optical transparency, high thermal conductivity, and large surface to volume ratio. Here we present the latest results on the enhancement of the infrared optical absorptance and ultraviolet photoluminescence in ZnO nanocolumns after hydrogen plasma treatment. The photoluminescence in near UV region at 378 nm has been significantly enhanced by hydrogen plasma treatment at roo temperature whereas the defect related yellow photoluminescence (broad band 550–650 nm) decreased.

1. Introduction

ZnO is an direct band gap semi-conductor with optical transparency in the visible spectral range, large exciton binding energy and related room temperature photoluminescence, tunable electrical conductivity and high thermal conductivity [1]. The native point defects [2] and oxygen vacancies [3] have been studied including their sensitivity to hydrogen. Hydrogen works as a shallow donor as shown by photoluminescence and photoconductivity measurements [4]. The hydrogen plasma treatment of single crystal ZnO passivates deep defects and enhances the band edge luminescence [5].

The layers of vertically aligned ZnO nanocolumns can be grown on glass substrate coated by thin polycrystalline films [6]. The nanostructure ZnO is attractive for applications in sensors and energy converters such as gas sensors [7] and biosensors [8] as they possess useful electrical and optical properties such as tunable electrical conductivity, low optical absorbance and room temperature luminescence [9]. The annealing in H_2 atmosphere at temperature above 800°C enhances significantly excitonic emission in ZnO nanoparticles [10]. The experimental results indicate that the OH and H bonds play the dominant role in facilitating surface recombination [11].

In our technology, the seeding layer is deposited by DC reactive magnetron sputtering of Zn target in the gas mixture of argon and oxygen plasma followed by the hydrothermal growth of ZnO nanocolumns. We have shown that the lack of the seeding layer leads to random growth of low quality nanocrystals [12]. We have found that the oxidation by the thermal annealing in air significantly reduces the near infrared optical absorption as well as the electrical conductivity in correlation with the increased presence of non-lattice oxygen in the form of Zn–O–H and Zn–O–Zn [13]. On the contrary, the increase of the infrared optical absorption and the electrical conductivity related to free carrier concentration was detected below the optical absorption edge after hydrogen plasma treatment

[14]. Our latest results showed a significant increase in the length of the ZnO nanocolumns grown under UV irradiation, as well as the improvement in their uniformity [15].

2. Experimental

2.1. Sample preparation

The seeding layers were prepared by reactive magnetron sputtering in the stainless-steel vacuum chamber using a Zn target and O_2 plasma. The Zn target was sputtered in a capacitive coupled dc glow discharge plasma (grounded substrate holder at temperature 400°C, constant potential of +400 V on the target, DC current of 0.13 A) of a reactive mixture of argon (purity 99.99%, flow rate 2.0 sccm) and oxygen (purity 99.95%, flow rate 0,5 sccm) under pressure of 1 Pa [16].

The ZnO nanorods grew on nucleated fused silica glass substrates by hydrothermal process in an oil bath containing a flask with ZnO nutrient solution that contained a mixture of two equimolar aqueous solutions of 25 mM zinc nitrate hexahydrate ($Zn(NO_3)_2.6H_2O$) and hexamethylenetetramine ($C_6H_{12}N_4$). The resulting solution was preliminary stirred at 60°C for 1 h and then filtered using a polytetrafluoroethylene (PTFE) filter with a pore size of 0.45 μm. Finally, the samples was carefully washed in deionized water and purged in nitrogen. During the ZnO nanostructural growth, the substrates were mounted upside-down on a flat bottom quartz tube with an optical fiber placed inside the tube. The UV LED light source operating at power 1 W and the wavelength 365 nm was turned on to perform the UV-assisted growth. The average height of ZnO nanorods was 1250 nm and the diameter 200 nm.

The plasma hydrogenation was done in the stainless-steel vacuum chamber at room temperature using 9 W rf discharge at 13.56 MHz, hydrogen flow 50 sccm and pressure 70 Pa. Prior the plasma hydrogenation, the vacuum chamber had been evacuated down to the pressure vacuum 10^{-7} Pa.

2.2. Photoluminescence spectroscopy

The steady state photoluminescence spectra (PL) of the highly scattering thin films were measured in the 360–660 nm spectral range using the f/4 double gratings monochromator SPEX equipped with two 1200 grooves/mm gratings blazed at 500 nm. The photoexcitation was provided by the 1 mW LEDs at the wavelength 340 nm and optically filtered by narrow band pass optical filter featuring 90% transmission and OD 6 blocking outside of the 20 nm passband. The sample holder was positioned by two perpendicularly oriented translation stages manually driven by adjuster screws for precision motion. The emitted light was collected and focused onto the 1 mm wide monochromator input slit by two 90° off-axis mirrors coated by UV enhanced aluminum. The signal was detected by the multi-dynode multi-alkali red sensitive photomultiplier (Photonis XP2203B) cooled to -12°C as the dc anode current from 1 pA up to 1 μA, coaxial cables and the electrometer Keithley 6517. The dark anode current at 1000 V was 2 pA and the noise 0.5 pA. The voltage was provided by the electrometer via coaxial cables. The PL spectra were measured at room temperature and calibrated on the spectral efficiency of the spectroflouorometer [17].

2.3. Photothermal deflection spectroscopy

The photothermal deflection spectroscopy (PDS) measures directly the optical absorption of thin films with sensitivity of four orders of magnitude [18]. The PDS spectrophotometer used the 150 W Xe lamp as a light source and the monochromator equipped with three gratings blazed at 300, 750, 1250 nm operating in a broad spectral range from ultraviolet to infrared region 250–1700 nm. The measured sample was immersed in transparent liquid with the probe laser beam passing parallel to the sample surface. The sample was illuminated by the monochromatic light. The heat absorbed in the sample deflected the probe laser beam detected by the position detector. Since the amplitude of the probe beam deflection was proportional to the optical absorption, the optical absorptance was measured by normalizing the signal on the deflection of the black sample.

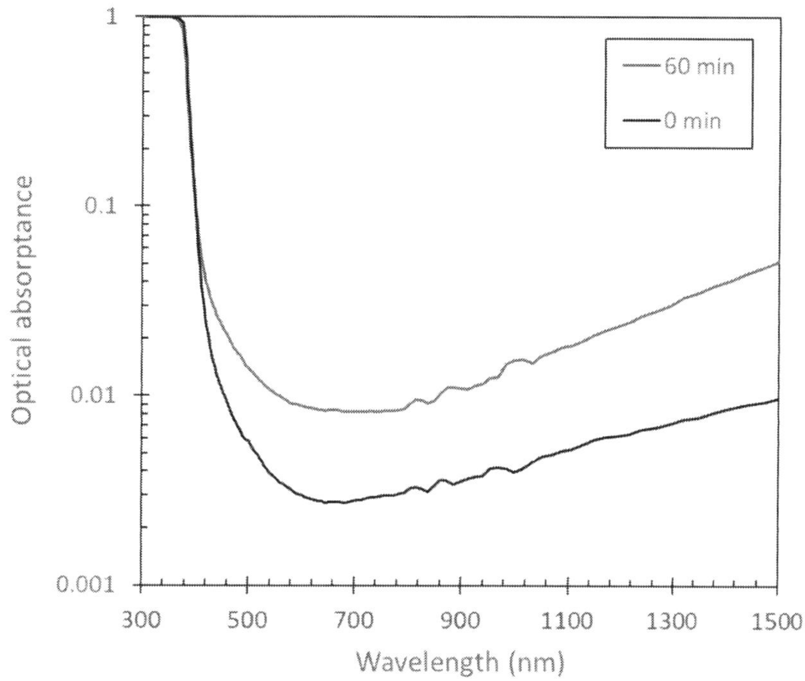

Figure 1. Optical absorptance spectra of ZnO nanocolumns before (0 min) and after plasma hydrogenation (60 min)

3. Results and discussion

Figure 1 shows the optical absorptance spectra of ZnO nanocolumns measured by the photothemal deflection before and after plasma hydrogenation. The optical absorption edge related to the ZnO band gap appears at wavelengths below 380 nm in both samples, The region of the free carrier absorptance appears in the near infrared region [19]. The infrared optical absorptance increased after hydrogen plasma treatment whereas the optical absorption edge has not been effected by plasma hydrogenation.

Figure 2 shows the photoluminescence emission spectra (PL) of the as grown and plasma hydrogenated fused silica glass substrate coated by the very thin nanocrystalline ZnO thin film by reactive magnetron sputtering. Since at 680 nm would appear the second harmonic of the excitation wavelength 340 nm, the emission spectra were measured only at wavelengths below 660 nm. On the other hand, the emission spectra were not measured at wavelengths below 360 nm because of the scattered excitation light. Figure 2 shows that the PL signal of the substrate is very low in the visible region above 400 nm being comparable with the background noise level. The signal increases in UV below 400 nm probably due to the scattered excitation light. However, there is a detectable weak emission peak at 378 nm that has been related to the indirect annihilation of intrinsic excitons with the simultaneous emission of one LO phonon [20]. Moreover, the emission related to deep defect states is visible in dark room by eye as a weak yellow luminescence in the spectral range 550–650 nm [21].

Figure 3 shows the enhanced photoluminescence emission spectra of the plasma hydrogenated ZnO nanocolumns compared to the as grown samples before the plasma hydrogenation. The strong emission peak appears centered at 378 nm. The broad defect related emission band in the spectral range 550–650 nm is reduced. The exciton-related emission increased by two orders of magnitude after plasma hydrogenation while the defect related emission band significantly decreased. The optimal plasma hydrogenation has been found to be about 30 min.

IOP Conf. Series: Materials Science and Engineering **465** (2019) 012008 doi:10.1088/1757-899X/465/1/012008

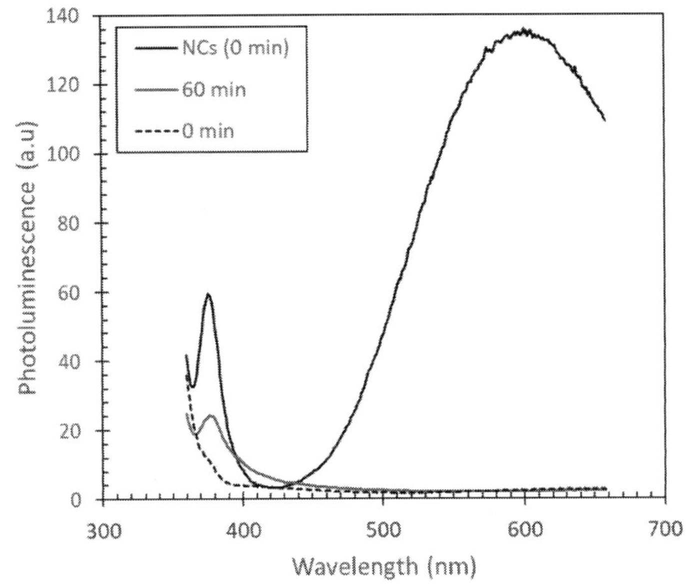

Figure 2. Photoluminescence spectra of the nucleated substrate before (0 min) and after plasma hydrogenation (60 min). The PL spectra of ZnO nanocolumns before the plasma hydrogenation (NCs 0 min) has been added for comparison and they are also shown in figure 3.

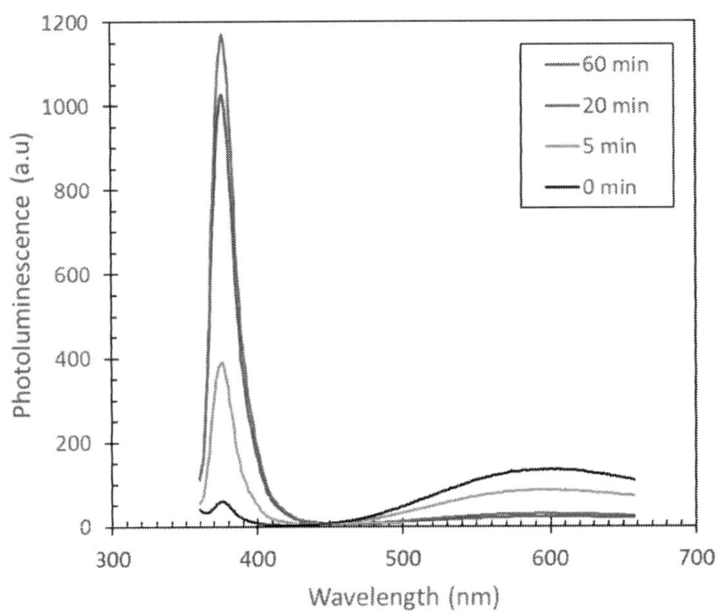

Figure 3. The enhanced photoluminescence emission spectra of the plasma hydrogenated ZnO nanocolumns compared to the as grown samples before the plasma hydrogenation.

Figure 4 compares the optical absorptance and photoluminescence spectra of the hydrogenated ZnO nanocolumns (UV enhanced growth, 60 min plasma hydrogenation). The horizontal axis has been recalculated to energy expressed in eV. Since the photoluminescence measurements has been done with constant spectral resolution 2 nm, the photoluminescence spectra expressed as a function of the energy needs to be normalized by multiplying with factor E^{-2}, where E is energy in eV [22]. This correction in conjunction with the conversion to line shape provides the properly treated emission data for quantitative analysis. Figure 4 shows that the photoluminescence peak has maximum at 3.3 eV just below the optical absorption edge.

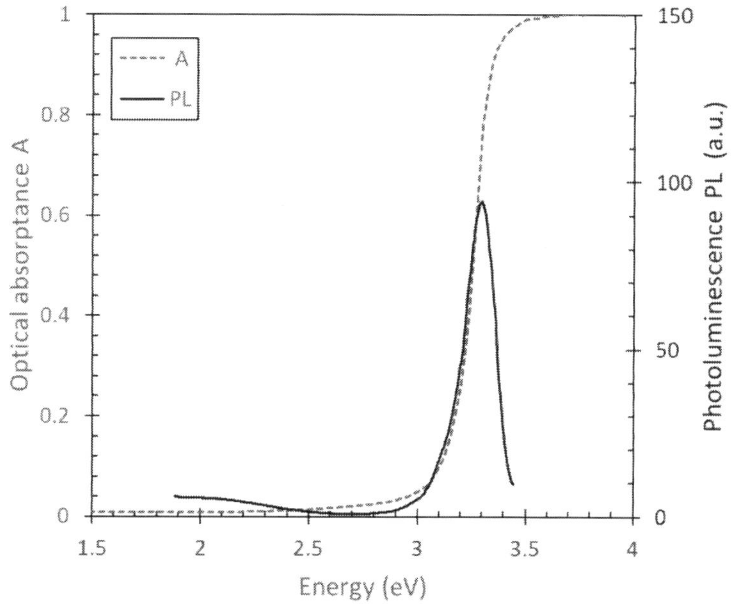

Figure 4. Comparison of the optical absorptance (A) and photoluminescence (PL) spectra of the hydrogenated ZnO nanocolumns (UV enhanced growth, 60 min plasma hydrogenation).

4. Conclusion

The optical absorption and photoluminescence emission spectra of thin films of the ZnO nanocolumns were measured using optical methods optimized for optically scattering samples. These methods include photothermal deflection spectroscopy and the photoluminescence spectroscopy measured using the double grating spectrofluorometer with very low background scattered light. The main advantages of our methods are the precise sample positioning of thin films, the high sensitivity and very low influence of the scattered light. This is very important for the highly scattering samples such as nanostructured thin films. We have observed that the exciton related emission band of ZnO nanocolumns centered at the wavelength 378 nm increased significantly after the hydrogen plasma treatment, whereas the defect related emission band at wavelengths 550–650 nm significantly decreased. These effects correlate with the increase of the infrared absorption. The optimal plasma hydrogenation has been found to be about 30 min.

Acknowledgements

This work was supported by the CSF project 16-10429J, the MEYS project LTC17029 INTER-COST

Action MP1406 and by the Operational Programme Research, Development and Education financed by European Structural and Investment Funds and the Czech Ministry of Education, Youth and Sports (Project No. SOLID21 - CZ.02.1.01/0.0/0.0/16_019/0000760).

References

[1] Janotti A and Van de Walle C G 2009 Fundamentals of zinc oxide as a semiconductor *Rep. Prog. Phys.* **72** 126501

[2] Janotti A and Van de Walle C G 2007 Native point defects in ZnO *Phys. Rev. B* **76** 165202

[3] Janotti A and Van de Walle C G 2005 Oxygen vacancies in ZnO *Appl. Phys. Lett.* **87** 122102

[4] Li Q H, Gao T, Wang Y G and Wang T H 2005 Adsorption and desorption of oxygen probed from ZnO nanowire films by photocurrent measurements *Appl. Phys. Lett.* **86** 123117

[5] Sekiguchi T, Ohashi N and Terada Y 1997 Effect of Hydrogenation on ZnO Luminescence *Jpn. J. Appl. Phys. Part 2 Lett.* **36** L289

[6] Greene L E, Law M, Goldberger J, Kim F, Johnson J C, Zhang Y, Saykally R J and Yang P 2003 Low-temperature wafer-scale production of ZnO nanowire arrays *Angew. Chem. - Int. Ed.* **42** 3031

[7] Kaneti Y V, Zhang Z, Yue J, Zakaria Q M D, Chen C, Jiang X and Yu A 2014 Crystal plane-dependent gas-sensing properties of zinc oxide nanostructures: experimental and theoretical studies *Phys Chem Chem Phys* **16** 11471

[8] Tereshchenko A, Bechelany M, Viter R, Khranovskyy V, Smyntyna V, Starodub N and Yakimova R 2016 Optical biosensors based on ZnO nanostructures: advantages and perspectives. A review *Sens. Actuators B Chem.* **229** 664

[9] Margueron S and Clarke D R 2014 The high temperature photoluminescence and optical absorption of undoped ZnO single crystals and thin films *J. Appl. Phys.* **116** 193101

[10] Procházková L, Gbur T, Čuba V, Jarý V and Nikl M 2015 Fabrication of highly efficient ZnO nanoscintillators *Opt. Mater.* **47** 67

[11] Yang L L, Zhao Q X, Willander M, Liu X J, Fahlman M and Yang J H 2010 Origin of the surface recombination centers in ZnO nanorods arrays by X-ray photoelectron spectroscopy *Appl. Surf. Sci.* **256** 3592

[12] Mičová J, Buryi M, Šimek D, Drahokoupil J, Neykova N, Chang Y-Y, Remeš Z, Pop-Georgievski O, Svoboda J and Im C 2018 Synthesis of zinc oxide nanostructures and comparison of their crystal quality *Appl. Surf. Sci* **461** 190

[13] Neykova N, Stuchlik J, Hruska K, Poruba A, Remes Z and Pop-Georgievski O 2017 Study of the surface properties of ZnO nanocolumns used for thin-film solar cells *Beilstein J. Nanotechnol.* **8** 446

[14] Chang Y-Y, Stuchlík J, Neykova N, Souček J and Remeš Z 2017 Optical properties of the plasma hydrogenated ZnO thin films *J. Electr. Eng.* **68** 70

[15] Neykova N, Chang Y-Y, Buryi M, Davydova M, Kucerkova R, Simek D, Remes Z and Pop-Georgievski O 2018 Study of ZnO nanorods grown under UV irradiation *Appl. Surf. Sci.* (in press)

[16] Remes Z, Stuchlik J, Purkrt A, Chang Y-Y, Jirasek V, Stenclova P, Prajzler V and Nekvindova P 2016 The intrinsic submicron ZnO thin films prepared by reactive magnetron sputtering *NANOCON 2016 - Conference Proceedings* 8th International Conference on Nanomaterials - Research and Application, Hotel Voronez I, Brno, Czech Republic, 19 - 21 October 2016 (Ostrava, Czech Republic: TANGER Ltd.) pp 36–41

[17] Remes Z, Neykova N, Potocky S, Chang Y-Y and Hsu H S 2018 The hydrogen plasma doping of ZnO thin films and nanoparticles *NANOCON 2017 - Conference Proceedings* 9th International Conference on Nanomaterials - Research and Application, Hotel Voronez I, Brno, Czech Republic, 18 - 20 October 2017 (Ostrava, Czech Republic: TANGER Ltd.) pp 57–62

[18] Remes Z, Babchenko O, Varga M, Stuchlik J, Jirasek V, Prajzler V, Nekvindova P and Kromka

A 2016 Preparation and optical properties of nanocrystalline diamond coatings for infrared planar waveguides *Thin Solid Films* **618** 130
[19] Chang Y-Y, Neykova N, Stuchlik J, Purkrt A and Remes Z 2016 Hydrogen plasma treatment of ZnO thin films *NANOCON 2016 - Conference Proceedings* 8th International Conference on Nanomaterials - Research and Application, Hotel Voronez Brno, Czech Republic, 19 - 21 October 2016 (Ostrava, Czech Republic: TANGER Ltd.) pp 161–5
[20] Weiher R L and Tait W C 1968 Contribution of Excitons to the Edge Luminescence in Zinc Oxide *Phys. Rev.* **166** 791
[21] Amiruddin R and Kumar M C S 2014 Enhanced visible emission from vertically aligned ZnO nanostructures by aqueous chemical growth process *J. Lumin.* **155** 149
[22] Mooney J and Kambhampati P 2013 Get the Basics Right: Jacobian Conversion of Wavelength and Energy Scales for Quantitative Analysis of Emission Spectra *J. Phys. Chem. Lett.* **4** 3316

IOP Conf. Series: Materials Science and Engineering **465** (2019) 012009 doi:10.1088/1757-899X/465/1/012009

Europium-doped Lu2O3 phosphors prepared by a sol-gel method

T Thoř[1], K Rubešová[1], V Jakeš[1], J Cajzl[1], L Nádherný[1], D Mikolášová[1], A Beitlerová[2] and M Nikl[2]

[1] Department of Inorganic Chemistry, University of Chemistry and Technology Prague, Technická 5, 166 28 Prague, Czech Republic
[2] Institute of Physics of the Czech Academy of Sciences, Cukrovarnická 10, 162 53 Prague 6, Czech Republic

Tomas.Thor@vscht.cz

Abstract. Cubic lutetium sesquioxide (Lu2O3) presents a suitable host material because of its favourable properties including chemical and phase stability. Lu2O3 has a wide band gap (> 5.5 eV) and its doping with lanthanide ions, which behave as activators, gives rise to some unique optical properties. Also, due to a high effective atomic number (Zeff = 68.8) and high density (9.42 g/cm3) of the Lu2O3 host, providing high stopping power for ionizing radiation, it has also attracted attention as a scintillator material. Since optical properties are affected by both crystal structure and morphology, which depend strongly on the synthesis method and conditions employed, many different methods have been studied for its synthesis. However, due to high melting temperature of Lu2O3 (2490 °C), the preparation of single crystals is difficult. In our work, we prepared polycrystalline samples of Eu:Lu2O3 using a sol-gel route based on the Pechini polycondensation. We tested three europium doping concentrations, two different annealing regimes were examined as well. The influence of mentioned parameters on photoluminescent and radioluminescent properties was established. Our samples proved that the used synthesis method can be used for further research of Lu2O3 doped with other rare-earth metals.

1. Introduction

Lutetium oxide is a phosphor material, whose emission properties can be tuned and enhanced by incorporating suitable dopants into its structure. As such it can find use in different lightning related applications, such as fluorescent lamps, cathode ray tubes or plasma display panels [1–3]. Furthermore, thanks to its high density of 9.42 g/cm^3 and high effective atomic number (Z_{eff} = 68.8), prompting high stopping power for ionizing radiation, it can also serve as a scintillating material (doped and even un-doped), converting high energy ionizing radiation to UV-Vis photons [4]. Lutetium oxide crystalizes in a bixbyite type cubic structure. Its crystal cell consists of a total of 32 cation sites, from which 24 show the low-symmetry C_2 point group and the remaining 8 show inversion symmetry having the C_{3i} (S_6) point group [5]. The inversion symmetry of the S_6 site leads to significantly lower cross section for dopant ions implemented on this site compared to those ions on C_2 sites. Therefore optical properties of doped Lu_2O_3 are dominated by dopant ions occupying low symmetry C_2 sites [6,7]. However, the way in which these sites are filled with dopant cations is not yet fully understood and therefore it is often approximated as a random process.

Lutetium oxide possesses a high melting point of 2490 °C. On one hand, this empirically points towards a high chemical and mechanical stability, on the other hand, it makes the fabrication very difficult [6]. Such extremely high temperature presents a significant challenge for single crystal growth processes. Therefore, other methods have been studied for its fabrication, such as precipitation [8–11], combustion synthesis [12], polymer complex solution method [13], flame spray pyrolysis [14] or the sol-gel route [15].

In this work, we used the Pechini-type sol-gel method to prepare a series of europium-doped lutetium oxides. Europium ions act as an activator in Lu_2O_3 host, having the energy levels in the band gap of Lu_2O_3 and exhibiting a characteristic emission in the visible region of electromagnetic spectrum. The concentration of europium was selected based on published optimal concentration of europium in a very similar yttrium oxide host [16,17]. Then we prepared three samples with the concentration interval around the optimized one: $Eu_xLu_{(2-x)}O_3$, where x = 0.1, 0.15 and 0.2, corresponding to 5%, 7.5% and 10% Eu^{3+} doping, respectively. We used two different regimes for the final heat treatment, 1000 °C/4 h and 1250 °C/24 h, and its influence on the studied luminescence properties was established.

2. Materials and methods

2.1. Bulk samples preparation

Lutetium oxide (Crytur Turnov; 99.99 % (REO)) was first dissolved in HNO3 (Penta; 65 %; p.a.) in a molar ratio of 1 mol Lu : 3.5 mol HNO3. Concentrated HNO3 was diluted with water in a volume ratio of 1:2. Mixture was stirred at 60 °C until clear solution was formed. Then europium acetate Eu(O2CCH3)3 (Strem chemicals; 99.9 % (REO)) was added to this solution. After dissolving, citric acid (CA; Lach:ner; anhydrous; p.a.) was added to the solution together with additional water required for a saturated solution of CA. The molar ratio of metal cations:CA was 1:4. Ethylen glycol (EG; Penta; 99 %) was added to the solution with the ratio of metal:EG being 1:16. The resulting solution was stirred at 80 °C until water was evaporated. Temperature was further increased to initiate polycondensation leading up to gelation. As prepared gel was dried at 250 °C in air for two hours. Dried xerogel was then subjected to a series of thermal treatments: first at 500 °C for 2 hours with 3 hours long ramp, then at 900 °C for 2 hours with 1.5 hours long ramp and rapid cooling at the end. Before the final heat treatment, powder samples were pressed into 1x2x10 mm pellets. The final annealing was performed in two regimes: 1000 °C for 4 hours and 1250 °C for 24 hours.

2.2. Characterization methods

X-ray powder diffraction (XRD) data were collected at room temperature with Bruker AXS D2 Phaser powder diffractometer with parafocusing Bragg-Brentano geometry using CoK_α radiation. Data evaluation was performed in the software package HighScore Plus (Malvern Panalytical).

The steady-state photoluminescence properties of the prepared samples were measured on a HORIBA Jobin Yvon Fluorolog®-3 Extreme spectrometer using FluorEssence™ 3 software. The spectra were collected at room temperature within the range of 450–800 nm. A photomultiplier tube (PMT) with thermoelectric cooling and a Ce:InGaAs photocathode (model number R955, detection range 185–900 nm) were used. Sample excitation was performed using a 450 W xenon continuous-wave (CW) lamp. To select various wavelengths, a double-diffraction-grating monochromator at the entrance and a single-diffraction-grating monochromator at the exit were used. The photoluminescence spectra were collected in a reflective arrangement with the sample being tilted at angle of approx. 60°. The photoluminescence radiation was collected at the front-facing exit. For the spectra evaluation, all of the measured luminescence spectra were transformed to the base level and, after subtraction of the background, normalised with the help of a reference sample (a single crystal of Ce:YAG).

The radioluminescence spectra were measured at room temperature. The custom made 5000M Horiba Jobin Yvon fluorescence spectrometer with a TBX-04 photon counting detector (IBH Scotland) was used. The samples were excited by an X-ray (40 kV,15mA) tube (SeifertGmbh). All spectra were corrected for the spectral distortions due to the setup. Spectral resolution used in the radioluminescence spectra measurements was 8 nm.

3. Results and discussion

Two series of europium-doped Lu_2O_3 were prepared using two different annealing temperatures (1000 °C or 1250 °C). Within each series, three samples with different europium doping were prepared - $Eu_xLu_{(2-x)}O_3$, where x = 0.1, 0.15 and 0.2, which corresponds to 5%, 7.5% and 10% Eu^{3+} doping, respectively.

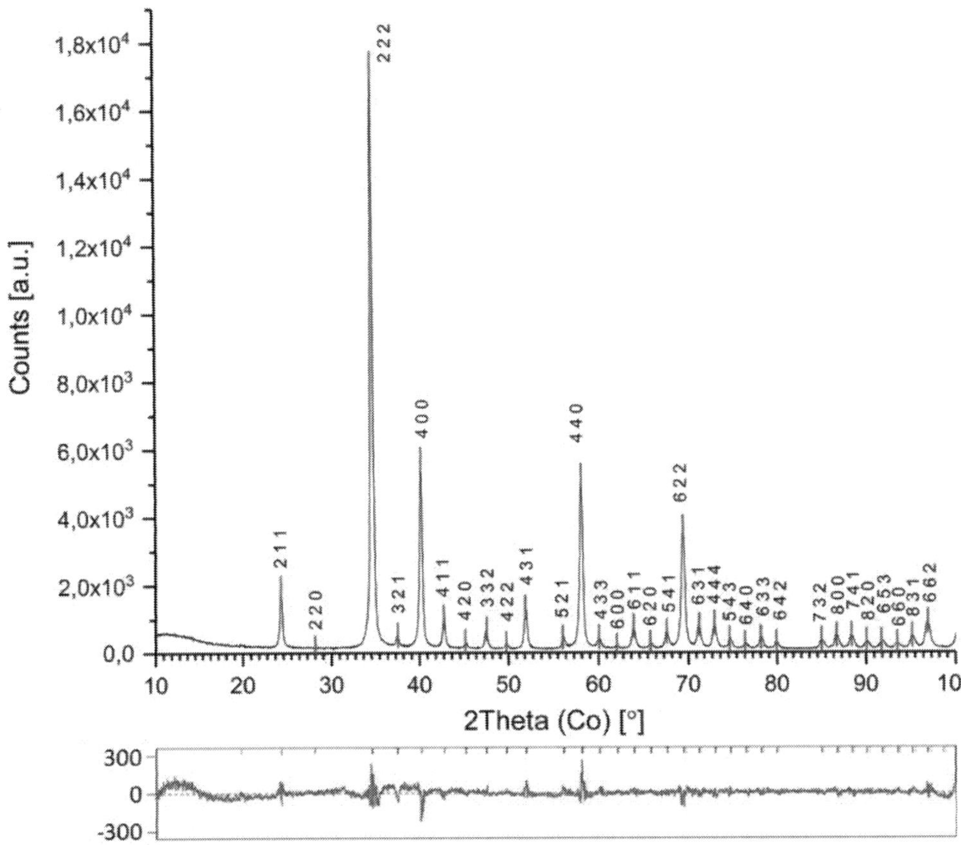

Figure 1: XRD pattern and difference plot of $Eu_{0.15}Lu_{1.85}O_3$ annealed at 1000 °C.

X-ray diffraction measurement showed that all of europium-doped Lu_2O_3 samples were single phase. As shown in the case of $Eu_{0.15}Lu_{1.85}O_3$ in figure 1, all of reflections in XRD pattern can be assigned to a cubic bixbyite type structure of un-doped Lu_2O_3, according to PDF 01-086-2475. Even for the highest concentration of europium used, no minor phase was detected. However, the europium doping influenced the unit cell size. After the Pawley fitting of XRD patterns, the unit cell parameter a was determined. In the scope of concentrations used, the dependence of lattice parameter on Eu concentration was observed to be linear (figure 2). Because of the effect of lanthanide contraction, ionic radii of europium ions are higher than in the case of lutetium ions, and thus substituting europium into a lutetium oxide host leads to an increase of lattice parameter. In terms of different annealing temperatures used, samples annealed at higher temperatures showed sharper diffraction peaks with smaller FWHM, which can be attributed to better crystallinity of those samples.

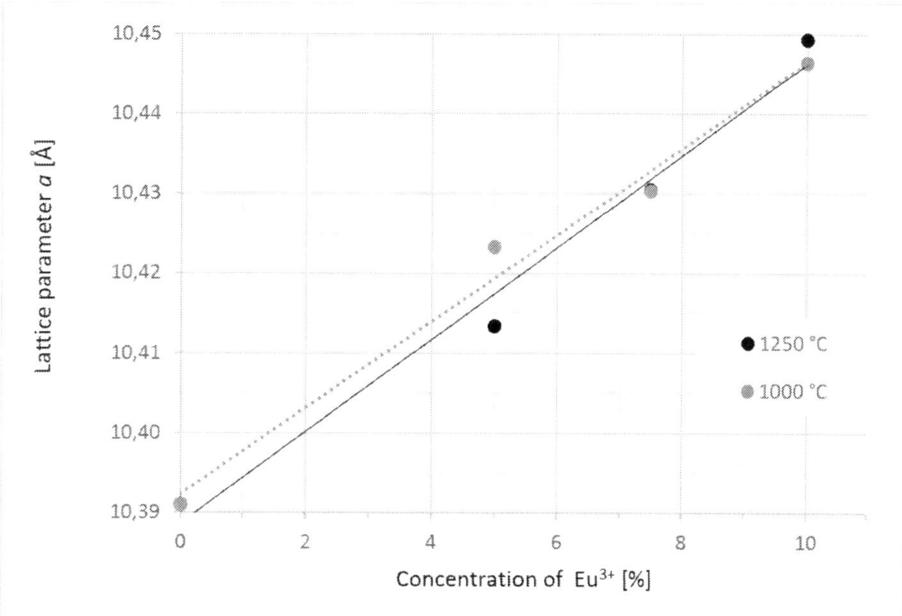

Figure 2. Lattice parameter of Eu:Lu2O3 as a function of increasing concentration of Eu3+ ions.

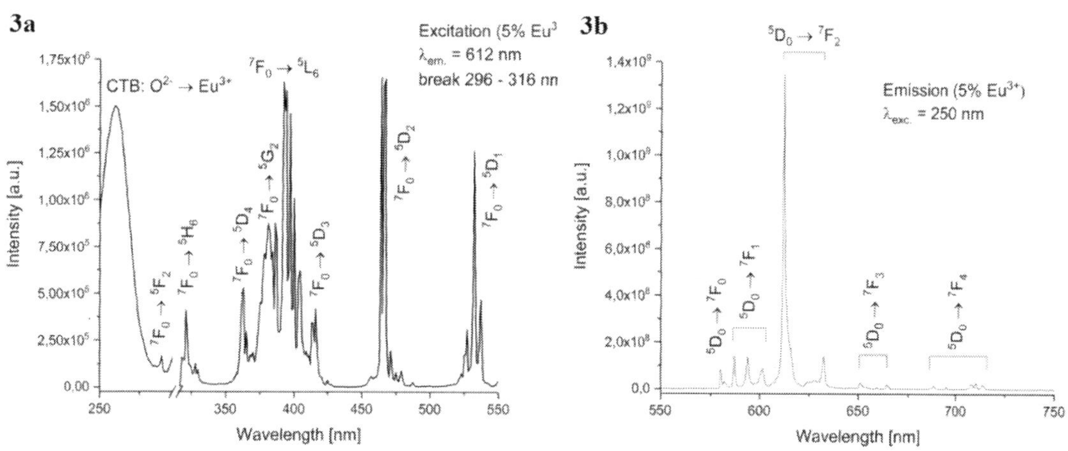

Figure 3. Photoluminescence excitation (3a) and emission (3b) spectra of $Eu_{0.1}Lu_{1.9}O_3$ annealed at 1250 °C.

The photoluminescence excitation spectrum of Eu:Lu2O3 (figure 3a) shows a broad band at around 260 nm corresponding to the charge transfer band from O2-(2p) to Eu3+(4f). Other bands shown correspond to 4f-4f transitions of Eu3+ from the 7F0 ground state to higher levels. Due to instrumentation issues, we could not filter harmonic excitation, therefore data around that were cut out using axis break.

The emission spectrum of Eu:Lu2O3 (figure 3b) reflects the transitions from the excited 5D0 level to the 7FJ (J = 0, 1, 2, 3, 4) levels of Eu3+. The most intense line, centred around 612 nm, corresponds to the hypersensitive transition to the 7F2 energy level. This emission is assigned to Eu3+ ions on the C2 sites in the Lu2O3 host [18]. However, all emission bands show narrow band width demonstrating homogeneous distribution of the dopant within the Lu2O3 host having good crystallinity.

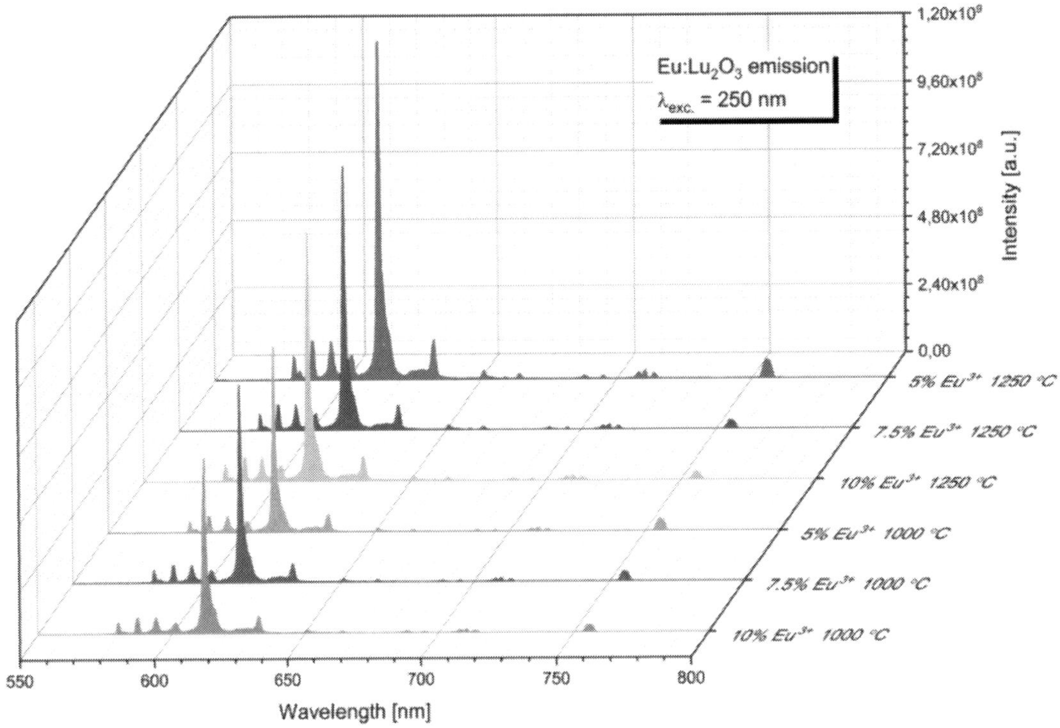

Figure 4: Photoluminescence emission spectra of $Eu_xLu_{(2-x)}O_3$ samples annealed at 1000 °C and 1250 °C.

Figure 4 shows photoluminescence emission spectra of samples with different dopant concentration and annealing temperature. Comparing overall photoluminescence emission intensity, it was higher in the case of samples annealed at higher temperature. This is most probably connected with the improvement of Lu2O3 host crystallinity and the better crystal vicinity of dopant ions. For the samples annealed at 1000 °C, the intensity increases slightly with dopant concentration from 5 to 7.5% and then slightly decreases for 10% Eu3+ concentration, most likely due to concentration quenching phenomenon caused by cross-relaxation between neighbouring Eu3+ ions. For the samples annealed at 1250 °C we can see different trend, since the highest intensity is observed for 5% of Eu3+ and then drops with the increasing concentration of Eu3+.

Although worse in resolution, radioluminescence spectra (figure 5) resemble closely photoluminescence emission spectra. Again, in all the samples, the emission pattern matches fully the emission of Eu3+ 4f-4f transitions, with the most intense one at 612 nm belonging to the 5D0 □ 7F2 energy transfer. For lower annealing temperature (1000 °C), the emission intensity increases monotonically with Eu concentration and further increases when the higher temperature of annealing was used. Within the 1250 °C series, the highest emission intensity was measured for 7.5% of Eu; however, the difference between 5% and 7.5% was negligible. 10% of Eu caused the decrease of the radioluminescence intensity. Despite the polycrystalline character of our samples, the measured radioluminescence intensity was by several orders of magnitude higher than for the reference BGO (bismuth germanate) single crystal (the red line in figure 5).

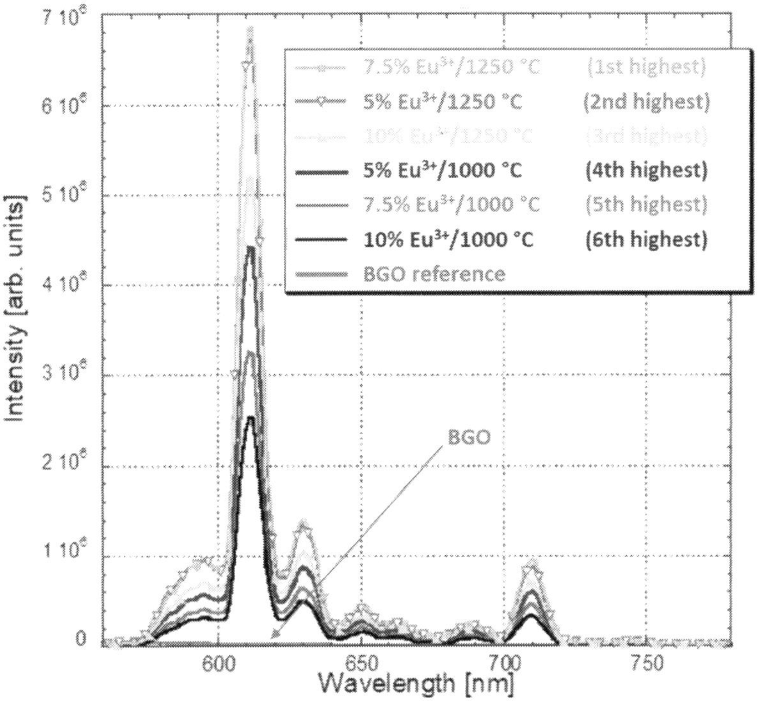

Figure 5. Radioluminescence spectrum of $Eu_xLu_{(2-x)}O_3$ (also the reference spectrum of BGO standard is included).

4. Conclusions

Lutetium oxide presents suitable host for the use either in phosphors or in scintillation applications. Here, we tested Eu-doped Lu_2O_3 prepared in a polycrystalline form. The dopant ions were successfully incorporated into Lu_2O_3 host for all concentrations used, not causing any phase changes. Photoluminescence and radioluminescence emission spectra were dominated by the f-f transitions of Eu^{3+} ions. Concentration quenching was clearly present for some doping concentrations and was probably caused by cross-relaxation processes in ion pairs. Annealing at higher temperature (1250 °C) increased both photoluminescence and radioluminescence emission intensities, most probably due to better crystallinity of the Lu_2O_3 host. The presented samples show that polycrystalline Lu_2O_3 can be successfully used for the further research of lanthanide dopants never previously examined.

Acknowledgment

This work was financially supported by specific university research (MSMT No 20-SVV/2017) and by the TA CR, project No. TJ0100041

References

[1] Yang J and Lin J 2010 Sol-Gel Synthesis of Nanocrystalline Yb3+/Ho3+-Doped Lu2O3 as an Efficient Green Phosphor *J. Electrochem. Soc.* **157** K273

[2] Boyer J C, Vetrone F, Capobianco J A, Speghini A and Bettinelli M 2004 Variation of Fluorescence Lifetimes and Judd-Ofelt Parameters between Eu3+ Doped Bulk and Nanocrystalline Cubic Lu2O3 *J. Phys. Chem. B* **108** 20137

[3] Zhang H, Chen J and Guo H 2010 Electrospinning synthesis and luminescent properties of

Lu2O3:Eu3+ nanofibers *J. Rare Earths* **28** 232

[4] Dujardin C, García Murillo A, Pedrini C, Madej C, Goutaudier C, Koch A, Petrosyan A G, Ovanesyan K L, Shirinyan G O and Weber M J 2000 Synthesis and scintillation properties of some dense X-ray phosphors *Proc. Int. Conf. Inorg. Scintill. Their Appl. SCINT99* 527

[5] Chen X, Xiang G, Chai F, Zhang Z, Shi S, Xu F and Zhao J 2016 An alternate spin-coating strategy toward high-quality polycrystalline thin Lu2O3: Eu3+ film fabrication *Surf. Coat. Technol.* **302** 523

[6] Kränkel C 2015 Rare-Earth-Doped Sesquioxides for Diode-Pumped High-Power Lasers in the 1-, 2-, and 3- μm Spectral Range *IEEE J. Sel. Top. Quantum Electron.* **21** 250

[7] Concas G, Spano G, Zych E and Trojan-Piegza J 2005 Nano- and microcrystalline Lu 2 O 3 :Eu phosphors: variations in occupancy of C 2 and S 6 sites by Eu 3+ ions *J. Phys. Condens. Matter* **17** 2597

[8] Lu B, Li J-G and Sakka Y 2013 Controlled processing of (Gd,Ln)2O3:Eu (Ln = Y, Lu) red phosphor particles and compositional effects on photoluminescence *Sci. Technol. Adv. Mater.* **14** 064202

[9] Dulina N A, Baumer V N, Danylenko M I, Mateyehenko P V, Tolrnachev A V, Vovk O M and Yavetskiy R P 2013 Effects of phase and chemical composition of precursor on structural and morphological properties of (Lu0.95Eu0.05)(2)O-3 nanopowders *Ceram. Int.* **39** 2397

[10] Marton Z, Bhandari H B, Brecher C, Miller S R, Singh B and Nagarkar V V 2013 Fabrication of High-Resolution Lu2O3:Eu X-Ray Scintillator by Physical Vapor Deposition *Ieee Trans. Nucl. Sci.* **60** 983

[11] Kopylov Y L, Kravchenko V B, Dulina N A, Lopin A V, Parkhomenko S V, Tolmachev A V, Yavetskiy R P and Zelenskaya O V 2013 Fabrication and characterization of Eu3+-doped Lu2O3 scintillation ceramics *Opt. Mater.* **35** 812

[12] Seferis I, Michail C, Valais I, Zeler J, Liaparinos P, Fountos G, Kalyvas N, David S, Stromatia F, Zych E, Kandarakis I and Panayiotakis G 2014 Light emission efficiency and imaging performance of Lu2O3:Eu nanophosphor under X-ray radiography conditions: Comparison with Gd2O2S:Eu *J. Lumin.* **151** 229–34

[13] Whiffen R M K, Antic Z, Speghini A, Brik M G, Bartova B, Bettinelli M and Dramicanin M D 2014 Structural and spectroscopic studies of Eu3+ doped Lu2O3-Gd2O3 solid solutions *Opt. Mater.* **36** 1083

[14] Seeley Z M, Dai Z R, Kuntz J D, Cherepy N J and Payne S A 2012 Phase stabilization in transparent Lu2O3:Eu ceramics by lattice expansion *Opt. Mater.* **35** 74

[15] de Jesus Morales-Ramirez A, Garcia-Murillo A, de Jesus Carrillo-Romo F, Garrido-Hernandez A and Garcia-Hernandez M 2015 Influence of annealing temperature on structural and optical properties of Lu2O3:Eu3+, Tb3+ transparent films *Mater. Res. Bull.* **70** 173

[16] Hao B V, Huy P T, Khiem T N, Ngan N T T and Duong P H 2009 Synthesis of Y2O3:Eu3+ micro- and nanophosphors by sol-gel process *J. Phys. Conf. Ser.* **187** 012074

[17] Alarcón-Flores G, García-Hipólito M, Aguilar-Frutis M, Carmona-Téllez S, Martinez-Martinez R, Campos-Arias M P, Zaleta-Alejandre E and Falcony C 2015 Synthesis and fabrication of Y2O3:Tb3+ and Y2O3:Eu3+ thin films for electroluminescent applications: Optical and structural characteristics *Mater. Chem. Phys.* **149** 34

[18] Pedroso C C S, Carvalho J M, Rodrigues L C V, Holsa J and Brito H F 2016 Rapid and Energy-Saving Microwave-Assisted Solid-State Synthesis of Pr3+-, Eu3+-, or Tb3+-Doped Lu2O3 Persistent Luminescence Materials *Acs Appl. Mater. Interfaces* **8** 19593

IOP Conf. Series: Materials Science and Engineering 465 (2019) 012010 doi:10.1088/1757-899X/465/1/012010

Analysis of laser beam welding of the S650MC high strength steel using numerical simulation

Š Vrtiel and M Behúlová

Slovak University of Technology in Bratislava, Faculty of Materials Science and Technology in Trnava, Ulica Jána Bottu 2781/25, 917 24 Trnava, Slovakia

stefan.vrtiel@stuba.sk, maria.behulova@stuba.sk

Abstract. Laser beam welding (LBW) is used in a wide range of industries, e. g. in the mechanical engineering, R&D, electronics, medical, aerospace or automotive industries. LBW technology is very popular not only for the well-known advantages comparing to another processes of fusion welding but also due to the fact that it can be easily integrated into the automated production lines. Modelling and numerical simulation of LBW processes provides a possibility to explore in detail technological principles and complex phenomena associated with this technology. The contribution is focused on the application of numerical simulation for the analysis and design of parameters for laser welding of S650MC high-strength steel sheets. For this purpose, simulation model of LBW process was developed and applied for numerical simulation of temperature fields in the program code ANSYS. A verification experiment was carried out using TruDisk 4002 disk laser. During the laser welding experiment, temperatures were measured using thermocouples of the K-type. The macrostructure of the produced weld joints was assessed by light microscopy. Measured temperatures and dimensions of welded joints were compared with results obtained by numerical simulation. Finally, using the verified simulation model, suitable parameters for laser welding of sheets from S650MC high-strength steel with the thickness of 4 mm were suggested.

1. Introduction

Development of advanced high-strength steels (AHSS) have been connected with the growing demands of designers and engineers for relatively cheap materials with special properties particularly high strength, ductility, toughness or fatigue properties. Advanced high-strength steels are complex, sophisticated materials, with carefully selected chemical compositions and multiphase microstructures resulting from precisely controlled heating and cooling processes [1].

Generally, the AHSS microstructure is composed of a relatively soft ferrite matrix in combination with the reinforcing phase (martensite, bainite, precipitates) [2-3]. The production of AHSS using unconventional methods of thermal and thermo-mechanical processing leads to the formation of unique microstructures and microstructural components whose exceptional mechanical properties can be impaired by exposure to elevated temperatures [4-6]. This temperature sensitivity of AHSS can be taken into account in selection of welding methods for joining components from these steels. The localized heat input associated with fusion welding processes can cause significant changes in the local microstructure, and hence affect mechanical properties of AHSS in weld joint area. During laser beam welding, the energy needed for material joining is focused to a very narrow zone, which results in minimal degradation of initial properties of the base material, low residual stresses and distortions [7-10]. Moreover, the laser beam can be precisely located and the welding speeds can be very fast what makes the laser welding very attractive for many industrial applications including high-efficient production in the automotive sector.

This paper is focused on the numerical simulation of temperature fields during laser beam welding of high-strength steels. The main aim of the research is to design the appropriate parameters for butt laser welding of 4 mm thick sheets made of S650MC high-strength steel using numerical simulation of the welding process in the ANSYS program code.

2. Simulation model

To analyse the temperature fields during formation of butt joints of high-strength steel sheets by disc laser with the maximum power of 4 kW, simulation model of laser beam welding process was developed, experimentally verified and applied to fit proper welding parameters.

Numerical simulation of temperature fields is based on the numerical solution of Fourier-Kirchhoff partial differential equation in the form (1)

$$\rho c_p \frac{\partial T}{\partial t} = \frac{\partial}{\partial x}\left[\lambda_x(T)\frac{\partial T}{\partial x}\right] + \frac{\partial}{\partial y}\left[\lambda_y(T)\frac{\partial T}{\partial y}\right] + \frac{\partial}{\partial z}\left[\lambda_z(T)\frac{\partial T}{\partial z}\right] + \dot{q}_v \quad [\mathrm{W.m^{-3}}] \qquad (1)$$

in which T is the temperature, ρ is the density, c_p is the specific heat, λ_x, λ_y and λ_z are the thermal conductivities in x, y and z directions, respectively and q_v is the volume heat source density, i. e. the heat generated per unit time in a unit volume [11]. Geometrical, physical, initial and boundary conditions are necessary to define to accomplish solution and numerical analysis.

2.1. *Geometrical and finite element model*

Geometrical model was suggested taking into account symmetry of the temperature fields during laser welding of high-strength steel sheets according to central plane of weld joint. Relatively small computation domain with dimensions of 15 mm × 30 mm × 4 mm (Figure 1) was considered to save computing time. On the other hand, this small-sized geometrical model was fully sufficient if only the temperature fields during laser welding were studied. Of course, in case of the stress-strain analysis, geometrical model should correspond to the real geometry of components to be welded.

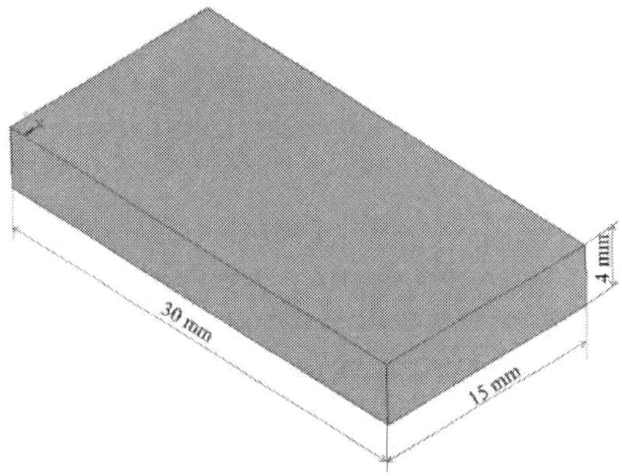

Figure 1. Geometrical model.

The finite element model (Figure 2) was generated in ANSYS, Release 18.1 [12] using 3D elements of the SOLID70 type. The size of elements (mesh density) was chosen taking into account different temperature gradients expected in the individual regions of the weld joint and the base material. The highest mesh density was near the weld centreline. The smallest element size was 0.2 mm in the welding direction and 0.05 mm in the direction perpendicular to the welding direction. In other zones of the weld joint and the base material, progressive meshing was used. The generated finite element model consists of 82122 nodes and 103144 elements.

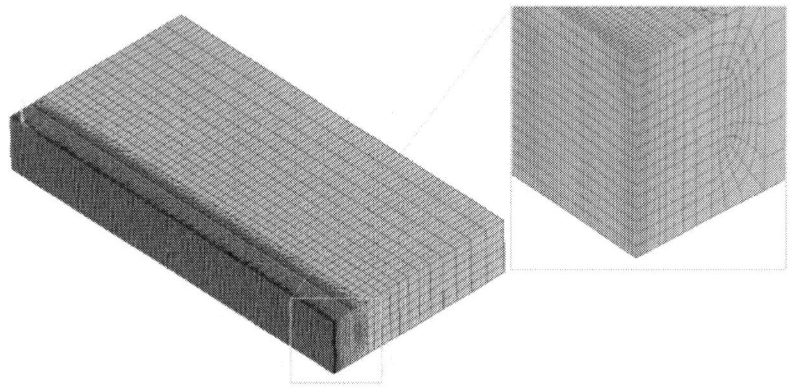

Figure 2. Finite-element model with a detail of the mesh density near the weld centreline.

2.2. *Material model*

Thermo-physical properties of the S650MC high-strength steel in the dependence on temperature were computed using the JMatPro software [13] on the base of chemical composition of the steel presented in Table. 1.

Table 1. Chemical composition of the S650MC steel.

Element	C	Si	Mn	P	S	Al	Nb	V	Ti	Mo	B	Fe
[wt. %]	0.12	0.60	2.00	0.025	0.015	0.015	0.09	0.20	0.22	0.50	0.005	bal.

Computed values of the thermal conductivity, density and specific heat of the S650MC steel *v. s.* temperature are plotted in Figures 3-5. For the numerical simulation of temperature fields during fusion welding, the enthalpy changes during the solid-state transformations as well as during the melting of the investigated steel are necessary to take into account. The amount of latent heat associated with single phase transformations was defined using the method of equivalent specific heat capacity [8]. The solidus and liquidus temperatures of the S650MC steel were defined on the base of computation to be 1414 ° C and 1500 ° C, respectively.

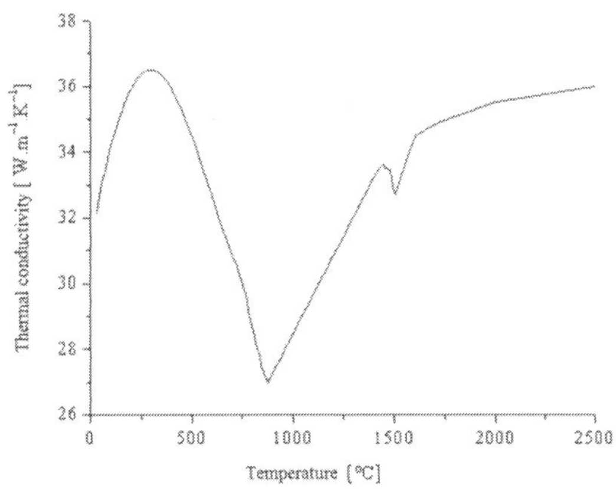

Figure 3. Thermal conductivity of the S650MC steel *vs.* temperature.

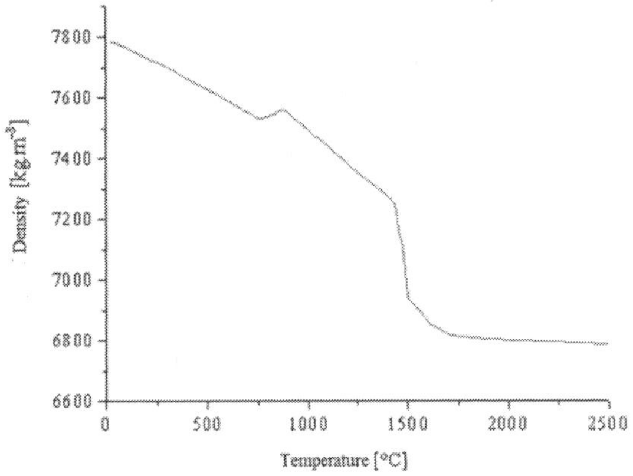

Figure 4. Density of the S650MC steel *vs.* temperature.

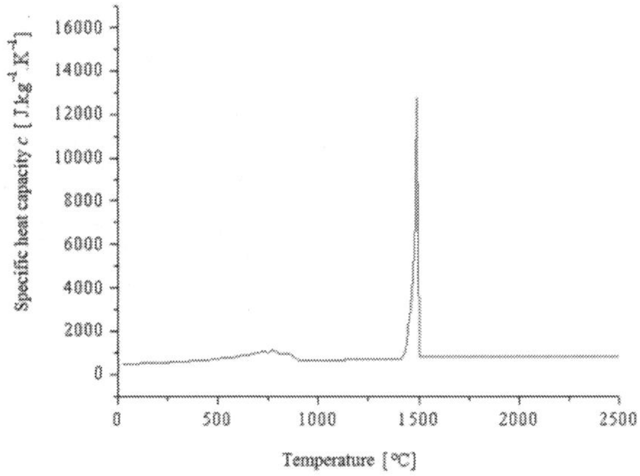

Figure 5. Specific heat of the S650MC steel *vs.* temperature.

2.3. Initial conditions, boundary conditions and loads

The initial temperature of the welded sheets was 20 °C. In performed numerical simulations, the heat dissipation from the samples to the environment was not considered due to the high welding speeds and consequently very short time periods for sample cooling during the welding process [14].

For the accurate computation of temperature field, substantially more significant is the modelling of the laser heat source [15]. In this case, the conical volumetric model of a heat source with Gaussian heat flow distribution was chosen (Figure 6). The conical volumetric heat source model is defined by the geometrical parameters of r_e, r_i, z_e and z_i according to Figure 6 together with the heat output of the laser beam source Φ and laser efficiency η [16].

$$q_v(x,y,z) = \frac{9\eta\Phi e^3}{\pi(e^3-1)} \times \frac{1}{(z_e - z_i)(r_e^2 + r_e r_i + r_i^2)} exp \frac{3(x^2+y^2)}{r_0^2(z)} \tag{2}$$

$$\text{where} \quad r_0(z) = r_e + \frac{r_i - r_e}{z_i - z_e}(z - z_e) \tag{3}$$

Based on previous experience [18], the following values of the parameters were applied in this simulation: $r_e = 0.079$ mm, $r_i = 0.072$ mm, $z_e = 4$ mm and $z_i = 0$ mm. The efficiency of the laser heat source was supposed to be 85 %.

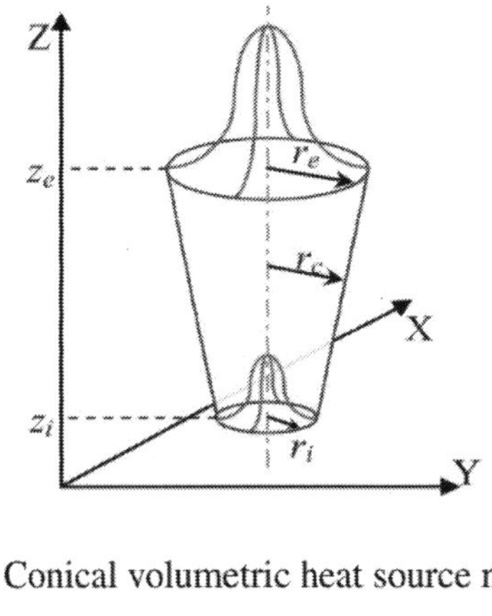

Figure 6. Conical volumetric heat source model [17].

Numerical simulation of temperature fields during laser welding of sheets made of the S650MC steel were performed using the ANSYS program code, Release 18.2 by means of the solution of a transient, non-linear thermal problem.

3. Verification experiment

To verify the developed simulation model, verification experiment was carried out in the Center of Excellence of the 5-axis machining at the Slovak University of Technology in Bratislava, Faculty of Materials Science and Technology (MTF SUT) in Trnava. The continuous wave TRUMPF TruDisk 4002 disc laser with the maximum power of 2 kW was used to prepare butt weld joints of samples from the S650MC steel sheet with dimensions of 50 mm × 100 mm × 4 mm. The wavelength of laser radiation was 1030 nm and the beam quality (BPP) 8 mm mrad. Laser beam was transported to the BEO D70 focusing optics via laser light cable with the core diameter of 200 µm. The focusing optics was mounted on the Fanuc M-710iC/50 6-axis robot. The disk laser beam was focused at +2 mm and +3 mm above the surface of the samples to be welded. Argon with the flow rates of 24 $L.min^{-1}$ and 30 $L.min^{-1}$ was used as a shielding gas. Both, weld bead and weld root were protected during laser beam welding. The laser power of 1500 W and 2000 W along with the welding speed from the range of 15 $mm.s^{-1}$ to 30 $mm.s^{-1}$ was exploited. Altogether, seven butt weld joints of samples were prepared experimentally using the welding parameters listed in detail in Table 2.

In the first trial, the laser welding with the laser power of 1500 W, the welding speed of 20 $mm.s^{-1}$and focusing +3 was performed. As this laser power was not sufficient to achieve the full weld root melting, the laser power was increased to 2000 W.

Figure 7. Laser workplace in the Center of Excellence at MTF SUT in Trnava.

Table 2. Parameters of experimental laser welding.

Sample number	Laser power P [W]	Welding speed w [$mm.s^{-1}$]	Focusing	Argon gas flow [$L.min^{-1}$]
1	1500	20	+3	24
2	2000	20	+3	24
3	2000	15	+2	30
4	2000	20	+2	30
5	2000	20	+2	30
6	2000	25	+2	30
7	2000	30	+2	30

According to visual evaluation of the prepared weld joint, the fusion and heat affected zones seemed to be relatively wide. In this reason, the focusing value was reduced to +2 and the next five weld joint were produced using the laser power of 2000 W and varying welding speeds.

During the welding process of the sample No. 5, experimental temperature measurement was carried out. Two thermocouples of the K-type were resistance welded to the experimental sample in the positions according to Figure 8. In this figure, the time dependence of temperatures measured by thermocouples during experiment is shown as well. The maximum temperature measured by the thermocouple 1 in the depth of 2 mm from the surface at the distance of 1.95 mm from the weld centreline is higher than the surface temperature measured by the thermocouple 2 distant from the weld centreline of 3.2 mm.

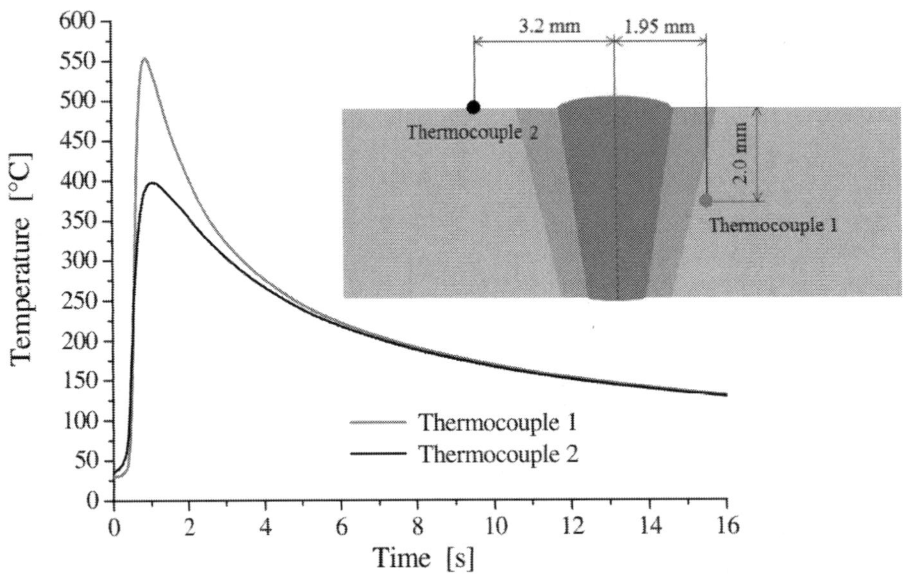

Figure 8. Location of thermocouples during experimental laser welding of the sample No. 5 and the time dependence of measured temperatures.

4. Comparison of results obtained by numerical simulation and verification experiment

Applying simulation model described in section 2, temperature fields during the laser welding with parameters provided in Table 2 were computed. The computed dimensions of fusion zones (weld pool) were compared with experimentally measured dimensions of fusion zones (FZ) and heat-affected zones (HAZ) on the macrostructures from the produced weld joints (samples No. 1 – 7). Moreover, for the sample 5, comparison of computed and experimentally measured temperatures was accomplished.

Figure 9 illustrates the temperature fields in the chosen times during the laser welding of the experimental sample No. 5. The fusion zone as well as HAZ are very narrow and the temperatures of the base metal distant from the weld centreline do not change during investigated time; they remain equal to the initial temperature of the sample. This indicate that the computation domain is large enough to evaluate the temperature fields during the studied process of laser welding. The statement can be confirm also by the Figure 10 illustrating the time-history of temperatures in central points at different distances from the welding start point. Already at the distance of 5 mm from the sample beginning, the welding process is in so-called regular mode, i. e. the computing domain supposed in the welding direction is sufficiently long.

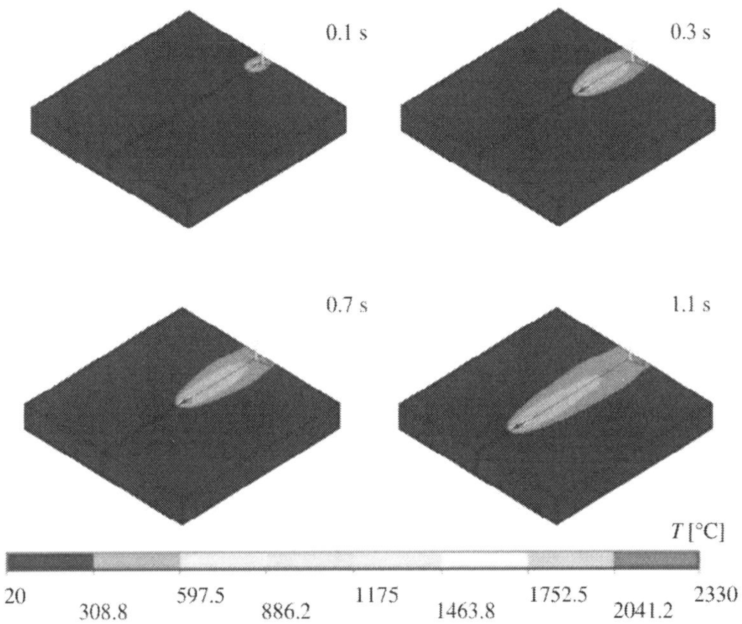

Figure 9. Temperature distribution in chosen times during the laser welding process (sample No. 5 - laser power of 2000 W, welding speed of 20 mm.s^{-1}).

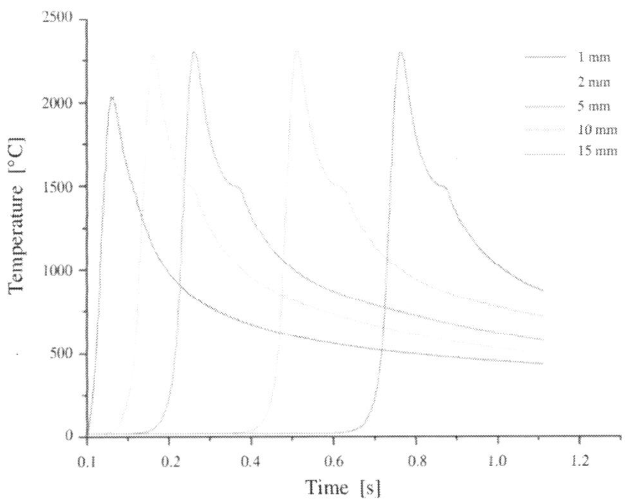

Figure 10. Time history of temperatures in central nodes at different distances from the welding start point.

In Figures 11-15, the temperature distribution and shapes of weld pool for weld joints No. 1, No. 3 and No. 5-7 are depicted using the sample top view, bottom view (or longitudinal cross-section view) and the 3D image of a fusion zone. The dark blue color represents material in solid state with temperatures below the solidus temperature of 1414 °C. In gray zones, temperatures are higher than the liquidus temperature for the S650MC steel (T_L = 1500 °C) and the material is fully melted. For comparison, macrostructures of prepared weld joints are presented in these figures as well. Macrostructural analysis was carried out using the Stemi 2000-C light microscope. The dimensions of

IOP Conf. Series: Materials Science and Engineering **465** (2019) 012010 doi:10.1088/1757-899X/465/1/012010

FZ and FZ+HAZ were evaluated according to the scheme in Figure 16. Table 3 summarizes the measured and computed dimensions of FZ and FZ+HAZ for all weld joint samples.

Sample No. 1 was produced with the laser power of 1500 W, focusing +3 and the welding speed of 20 mm.s^{-1} (Table 2). In this case, the base sheets with the thickness of 4 mm were not completely melted, i. e. incomplete root fusion occurred. The base material was melted only to the depth of 2.78 mm while the width of fusion zone at the top side was 1.50 mm. According to computation, the depth of fusion zone was 2.87 mm, the width and the length of fusion zone were 1.56 mm and 4.0 mm, respectively (Table 3), corresponding to the relative error at the level from 3 % to 4 %.

When the laser power was increased to 2000 W, the weld pool was relatively wide at the top side but too narrow at the bottom side (only 0.24 mm). After the focus adjustment, further weld joints were prepared with the maximum possible laser power of 2000 W and different welding speeds. With increased welding speed, the dimensions of FZ and HAZ decreases. Incomplete root penetration was achieved using the welding speed of 30 mm.s^{-1} (Figure 15).

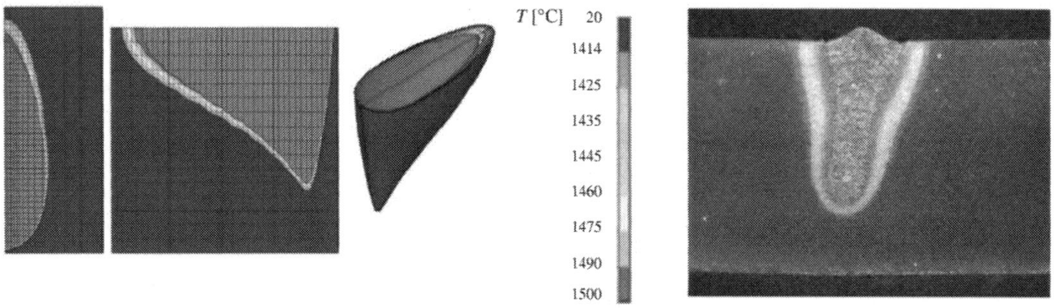

Figure 11. Temperature distribution during the laser welding of the sample No. 1 (top, bottom and 3D view of the weld pool) and macrostructure of the weld joint (P = 1500 W, w = 20 mm.s^{-1}).

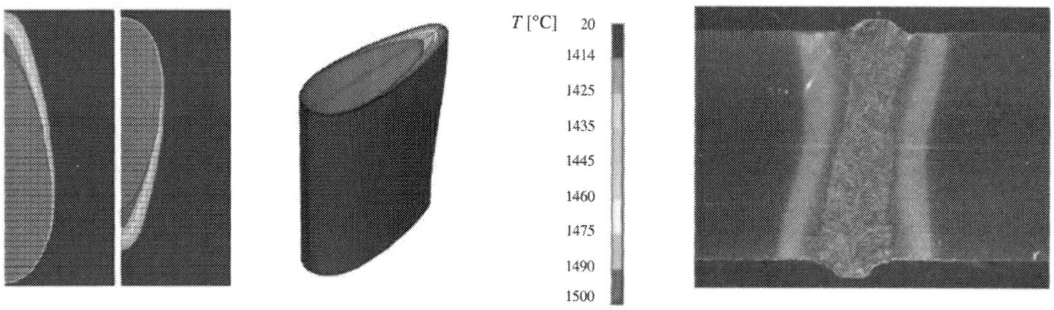

Figure 12. Temperature distribution during the laser welding of the sample No. 3 (top, bottom and 3D view of the weld pool) and macrostructure of the weld joint (P = 2000 W, w = 15 mm.s^{-1}).

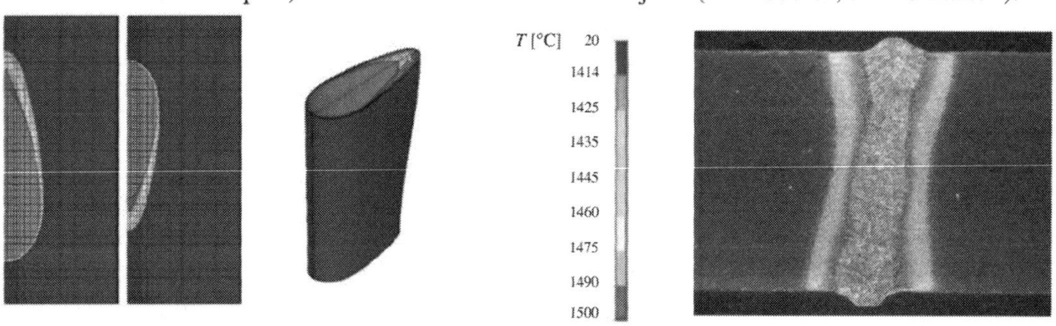

Figure 13. Temperature distribution during the laser welding of the sample No. 5 (top, bottom and 3D view of the weld pool) and macrostructure of the weld joint (P = 2000 W, w = 20 mm.s^{-1}).

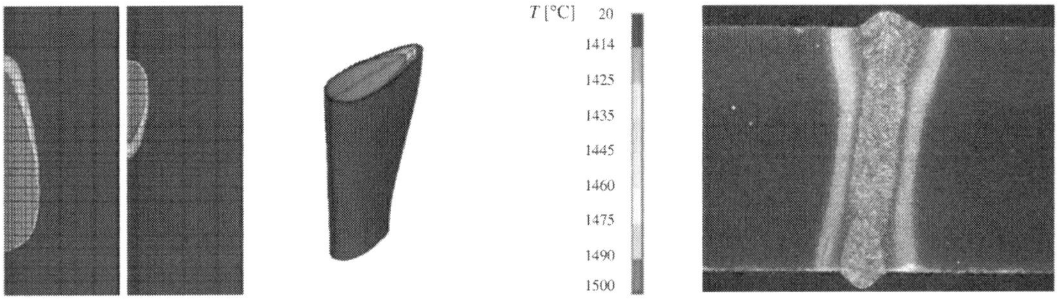

Figure 14. Temperature distribution during the laser welding of the sample No. 6 (top, bottom and 3D view of the weld pool) and macrostructure of the weld joint (P = 2000 W, w = 25 mm.s^{-1}).

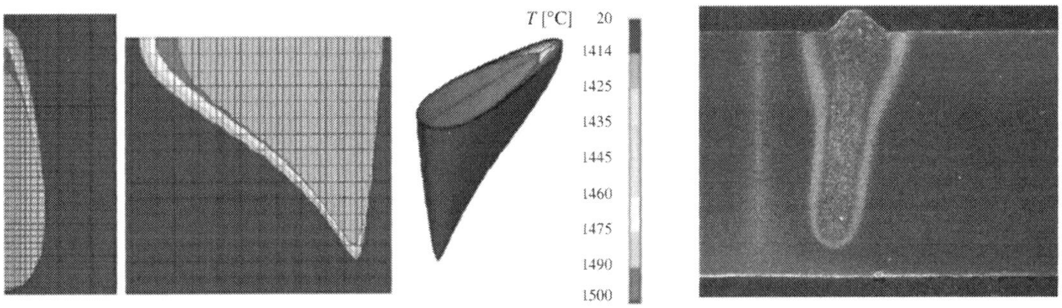

Figure 15. Temperature distribution during the laser welding of the sample No. 7 (top, bottom and 3D view of the weld pool) and macrostructure of the weld joint (P = 2000 W, w = 30 mm.s^{-1}).

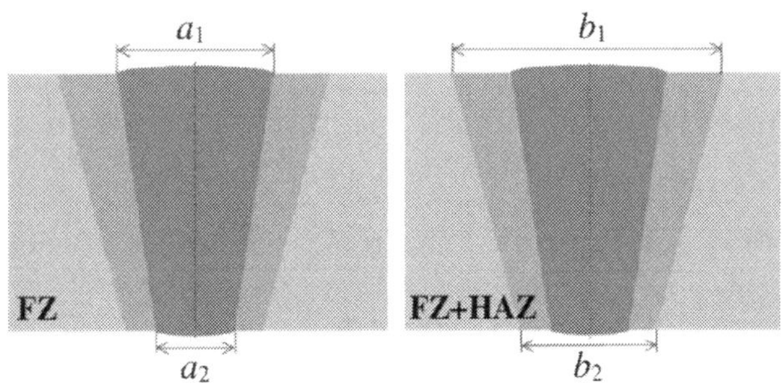

Figure 16. Scheme of the experimental measurement of the width of FZ and FZ+HAZ.

Table 3. Comparison of experimentally measured and calculated dimensions FZ and FZ+HAZ.

Sample No	P [W]	w [mm.s^{-1}]	Focus	FZ				FZ+HAZ	
				a_1 [mm]		a_2 [mm]		b_1 [mm]	b_2 [mm]
				Exp.	Calc.	Exp.	Calc.	Exp.	Exp.
1	1500	20	+3	1.50	1.56	**2.78***	**2.87***	2.31	2.87
2	1500	20	+3	1.53	-	0.24	-	2.58	-
3	2000	15	+2	1.42	1.60	1.52	1.46	2.69	1.46
4	2000	20	+2	1.09	1.26	1.11	1.06	2.2	1.06
5	2000	20	+2	1.26	1.26	1.12	1.06	2.33	1.06
6	2000	25	+2	1.17	1.14	0.81	0.66	2.08	0.66
7	2000	30	+2	1.12	1.28	**3.42***	**3.35***	1.86	3.35

* *penetration depth (full penetration was not achieved)*

In Figure 17, correlation between temperatures experimentally measured during the preparation of weld joint No. 5 and temperatures computed in nodes approximately corresponding to the position of thermocouples is shown. Considering the mesh generation and applied numerical method (FEM), it was not possible to obtain the time dependencies of calculated temperatures exactly in nodes of the thermocouple localization, therefore the compliance of measured and calculated temperatures can be considered as good. As it can be seen from calculated results, the temperature differences near the weld pool are significant even at very small distances, e. g. in the depth of 2 mm under the surface, the maximum temperature reached during the laser welding drops by more than 150 °C at the distance of 0.2 mm. Faster temperature decrease measured by thermocouple 2 in the depth of 2 mm can be caused by the heat extraction from the bottom surface of the welded sample to the fixture which was not considered in the numerical simulation.

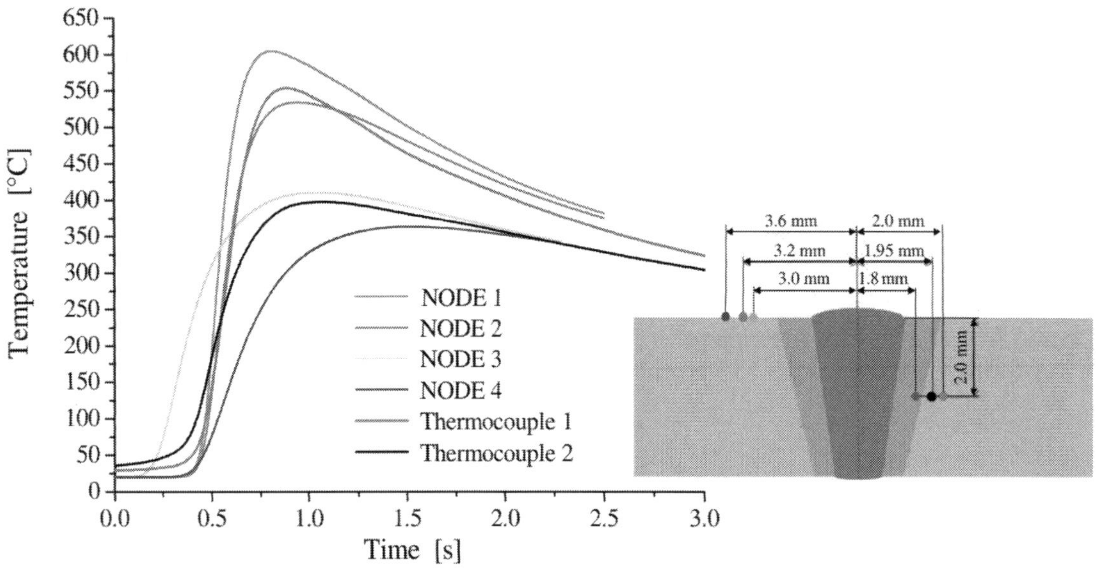

Figure 17. Comparison of time dependences of the temperatures measured during laser welding of the sample No. 5 and the temperatures calculated in selected nodes.

Based on the results of verification experiment, it can be concluded that the simulation model has been developed correctly and it can be used for further numerical analysis of the S650MC high-strength steel welding process.

5. Design of welding parameters for the laser welding of S650MC high-strength steel sheets

Design of appropriate welding parameters for production of high quality butt weld joints of S650MC high-strength steel sheets was carried out using numerical simulation with required laser power of 4000 W, focusing +2 and the welding speeds from the interval from 50 mm.s^{-1} to 100 mm.s^{-1}. The laser efficiency was supposed to be 85 %.

Figures 18 and 19 illustrate the shapes and dimensions of FZ computed in numerical experiments 1 – 6 for considered welding speeds by means of 3D views of molten zones and temperature fields in the cross-sections perpendicular to the welding direction. In Table 4, dimensions of FZ for welding parameters taken into account in numerical simulations are summarized together with computed maximal weld pool temperature. The increase in welding speed results in the decrease of FZ volume. Moreover, for the welding speed of 100 mm.s^{-1}, the laser power of 4000 W is not sufficient for complete root fusion.

Based on the results of numerical simulation, the welding speed of 70 mm.s^{-1} along with the laser power of 4000 W can be recommended for production of sound weld joints of 4 mm thick sheets made of the S650MC high-strength steel. This welding parameters can provide narrow weld joint with comparable weld width over the whole thickness of welded sheets. In addition, the weld pool temperatures are homogeneous. The computed maximal temperature of the weld pool is the lowest at the welding speed of 70 mm.s^{-1} (Table 4).

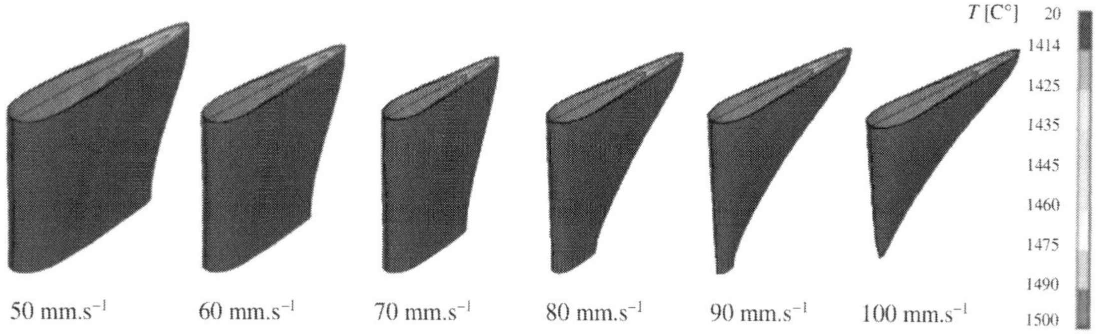

Figure 18. 3D view of fusion zones for welding speeds considered in numerical simulations.

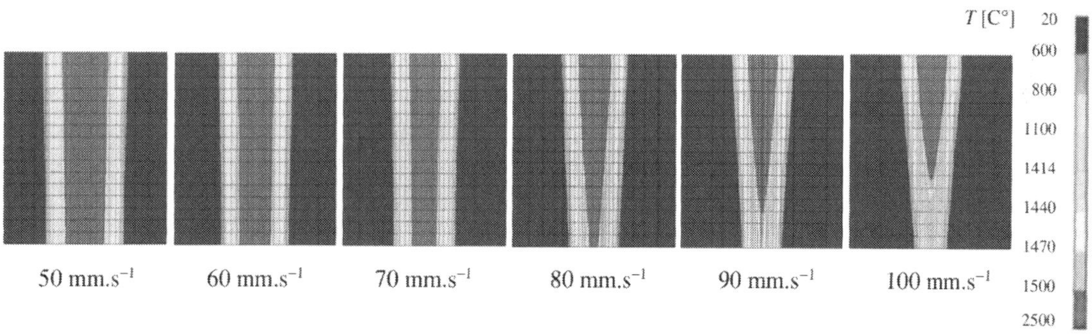

Figure 19. Temperature distribution in the weld joint cross-sections perpendicular to the welding direction for welding speeds from the range from 50 mm.s^{-1} to 100 mm.s^{-1}.

Table 4. Dimensions of fusion zone and the maximum temperature of a weld pool for welding parameters considered in numerical simulations.

Experiment No.	*P* [W]	*w* [mm.s^{-1}]	FZ		T_{max} [°C]
			a_1 [mm]	a_2 [mm]	
1	4000	50	0.54	0.46	2520
2		60	0.45	0.38	2200
3		70	0.38	0.30	1970
4		80	0.38	0.20	2050
5		90	0.38	0.10	2070
6		100	0.38	-	2090

6. Conclusions

The paper deals with numerical simulation of the laser welding of S650MC high-strength steel sheets with the thickness of 4 mm. The developed simulation model includes geometrical model with dimensions of 15 mm × 30 mm × 4 mm and finite element model generated in ANSYS program code using SOLID70 element type. Applying the JMatPro software, the material properties of S650MC steel in the dependence on temperature were calculated. A conical volumetric heat source was chosen to model the heat input during the laser welding. The experiment in the Center of Excellence MTF SUT in Trnava was carried out to verify the developed simulation model. Seven butt weld joints were produced using TruDisk 4002 disk laser with the maximum power of 2000 W and various welding parameters. During experiments, temperatures were measured by two thermocouples and consequently compared with results obtained by numerical simulations. The quality assessment of prepared weld joints was accomplished based on the macrostructure evaluation of welded samples and the measured dimensions of the FZ and HAZ. Finally, six numerical experiments (simulations) were performed using the developed and verified simulation model to design appropriate welding speed for the laser welding of S650MC high-strength steel sheets by a laser with the power of 4000 W. Based on the results of numerical simulations, the welding speed of 70 $mm.s^{-1}$ can be recommended to produce required butt weld joints.

Acknowledgments

The research has been supported by the Scientific Grant Agency of the Slovak Republic within the project VEGA No. 1/1010/16, KEGA No. 029STU-4/2018 and ITMS 26220120048 project under the Research & Development Operational Programme funded by the ERDF.

References

[1] Keller S, Kimchi M and Mooney P J 2017 *Advanced High-Strenght Steels Aplication Guidelines Version 6.0.* WorldAutoSteel, April 2017 Available at: https://www.world auto steel.org/projects/advanced-high-strength-steel-application-guidelines/ (Accessed 7.5.2018)
[2] Balogh A, Török I, Gáspár M and Juhász D 2012 *Production Processes and Systems* **5** 1 79-90
[3] Demeri M. Y. 2013 *Advanced High-Strength Steels—Science, Technology and Application* (ASM International, Ohio) ISBN-10: 1-62708-005-8
[4] Kuziak R, Kawalla S and Waengler S 2008 *Arch Civ Mech Eng* **8** 2 103-117
[5] Zhao J and Jiang Z 2018 *Prog Mater Sci* **94** 174-242
[6] Bublíková D, Jeníček Š, Vorel I and Mašek B 2017 *IOP Conf Ser: Mater Sci Eng* **179** 012009
[7] Biro E et al. 2010 *Metall and Mat Trans A* **41** 2348-56
[8] Shome M and Tumuluru M 2015 Introduction to welding and joining of advanced high-strength steels (AHSS). *Welding and Joining of Advanced High Strength Steels (AHSS)* (Amsterdam: Elsevier/Woodhead Publishing) chapter 1 1–8 (https://doi.org/10.1016/B978-0-85709-436-0.00001-1)
[9] Moeini G et al. 2018 *Mater Sci Eng A* **730** 232-243
[10] Kah P, Pirinen M, Suoranta R and Martikainen J 2014 *Advan Mater Res* **849** pp 357-365
[11] Incropera P F and De Witt D P 1996 *Fundamentals of heat and mass transfer* (New York: J. Wiley and Sons)
[12] ANSYS® Academic Research Mechanical, Release 18.1
[13] JMatPro Help, Release 6.1 2012 Sente Software Ltd
[14] Taraba B and Behúlová M 2007 Temperature fields in welding – Impact of modeling of weld joint formation and the heat transfer to the environment. *Proc. of the Meeting of Departments of Thermodynamic and Heat Transfer* (ZČU Pilsen) Available at: http://www.fluids.fs.cvut.cz/akce/konference/setkani_2007/sbornik/doc/Taraba.pdf) (Accessed 10.5.2018)
[15] Chukkan R J et al 2015 *J Mater Process Tech* **219** 48-59

[16] Zain ul Abdein M 2009 *Experimental investigation and numerical simulation of laser beam welding induced residual stresses and distortion in AA6056-T4 Sheets for Aeronautic application* (PhD Thesis) LaMCoS, INSA-Lyon.
[17] Zain ul Abdein M 2011 *Int J Pres Ves Pip* **88** 1 pp 45-56
[18] Bitala F 2018 *Analysis of laser welding of high-strength steels using numerical simulation* (Diploma Thesis) Slovak University of Technology in Bratislava (MTF SUT)

Corrosion resistance of electropolished AISI 304 stainless steel in dependence of temperature

V Zatkalíková and L Markovičová

University of Žilina, Faculty of Mechanical Engineering, Department of Material Engineering, Univerzitná 8215/1, 01026 Žilina, Slovakia

viera.zatkalikova@fstroj.uniza.sk

Abstract. AISI 304 is Cr-Ni austenitic stainless steel currently used for numerous applications that require high corrosion resistance. Due to the passive surface film this steel is resistant against a uniform corrosion, but halides present in environment can evoke local pitting. Electropolishing is recommended to obtain the high quality passive film and to increase the pitting corrosion resistance. This paper is focused on the assessment of the effect of elevated temperature (50 °C) on the resistance of AISI 304 electropolished surface to pitting corrosion in acidified 1 M chloride solution. Evaluation is based on the results of electrochemical cyclic potentiodynamic polarization tests (determination and comparison of pitting potentials E_p) and on the microscopic observation of specimen surfaces attacked by pitting during performed testing procedure.

1. Introduction

AISI 304 is the most widely used Cr-Ni austenitic stainless steel. It is recommended for a wide variety of applications that require high corrosion resistance, good formability and weldability. Common uses for this stainless steel type are found in such areas as food industry (e. g. cooking appliances, stoves, refrigerators, milk processing, wine-making, storage tanks), petrochemical, chemical and nuclear industry, dyeing industry, architecture and medicine. Other areas of use include chemical containers, heat exchangers, mining equipment, and marine nuts, bolts and screws, filtration systems.

Due to the passive surface film the stainless steels are resistant to the uniform corrosion in oxidation environments [1-4]. However, a presence of aggressive substances in environment can evoke local breakdown of protective film and consequently destructive and dangerous local corrosion, namely pitting [1, 2, 4-5].

There are a lot of various factors that influence the resistance of stainless steels to pitting. In addition to internal factors (chemical composition, surface treatment which affects the quality of passive surface film) important role is played by the environmental conditions (temperature, pH, concentration of aggressive ions) [1, 2, 5-7]. Temperature is a factor that strongly affects both thermodynamics and kinetics of pitting corrosion [1, 2, 7]. Numerous studies on the effect of temperature on pitting have been carried out. Most of them deal with the temperature in range 20 – 100 °C [1]. The authors [1, 2, 5-7] consider influence of temperature on pitting corrosion by the change of electrochemical characteristics – decrease of the pitting potential E_p and the repassivation potential E_r with temperature.

Many authors, e. g. [8-11] have documented an effect of surface characteristics on the pitting corrosion resistance of austenitic stainless steels. To obtain a high quality passive film which results in smooth material surface, various polishing technics are commonly used. Traditional mechanical polishing usually causes deformed layer and residual stresses on the treated surface [10]. Electropolishing as a surface finishing process of electrochemical anodic dissolution which enhances high surface brightness (mirror finish) without residual surface tensions appears as a better alternative. A particular layer from the metal's surface is removed and a high quality passive, chromium enriched oxide film is created [11]. According to the authors [12, 13], the increase of Cr/Fe ratio is caused by selective dissolution leading to the formation of a passive layer during the electropolishing process. Improved corrosion resistance and reduced bacterial attachment are other benefits of electropolished surfaces important for various industrial and medical applications of these materials [10].

This paper is focused on the pitting corrosion resistance of AISI 304 in acidified 1 M chloride solution. Resistance of three surface types ("as received", electropolished and ground + electropolished) is studied at compared at two temperatures (22 ± 3 °C, 50 °C). Evaluation of the corrosion resistance is based on the results of electrochemical cyclic potentiodynamic polarization tests and on the microscopic observation of specimen surfaces attacked by pitting during performed testing procedure.

2. Experimental material

The experimental material AISI 304 stainless steel (table 1) was purchased in sheet (1000 x 2000 mm) of 1.5 mm thickness. Its heat treatment was based on cold rolling, homogenization annealing (1040 – 1100 °C) and pickling after slightly smoothing rolling. The IIB surface finish means smooth and matte metallic glossy surface of the "as received steel" [14].

Table 1. Chemical composition of AISI 304 stainless steel (wt. %).

Cr	Ni	Mo	Mn	N	C	Si	P	S	Fe
18	8.01	-	1.40	0.075	0.027	0.38	0.031	0.0037	balance

Microstructure of experimental material (figure 1) is polyedric, created by austenitic grains with numerous twins, which could be created by annealing or by rolling. Parallel lines arose by the rolling during the technologic process.

Figure 1. Microstructure of AISI 304 stainless steel, cross section (glycerine + HNO_3 + HCl etch.).

3. Experimental procedure

Three types of studied steel surfaces were used for cyclic potentiodynamic tests (figure 2):

1. "as received" surface (AR), i.e. without additional mechanical /chemical treatment;
2. electropolished surface (EP);
3. ground (1200 mesh) and consequently electropolished surface (G+EP).

Electropolishing was performed in H_3PO_4 + H_2SO_4 + H_2O electrolyte under the conditions listed in table 2 [15, 16]. During electropolishing process the specimen was immersed into electrolyte and connected to the positive pole of the power source as an anode. A cathode was of the same material as the anode [16].

Figure 2. Types of steel surfaces used for potentiodynamic tests.

1M chloride acidified solution was used as the corrosion environment for electrochemical cyclic potentiodynamic test. This solution was represented by 0.9M NaCl + 0.1M HCl (pH = 1.1, specific conductivity 98 $mS.cm^{-1}$, redox potential 0.509 V). The used solution was chosen because of its similar aggressiveness and the same chloride concentration as the $FeCl_3$ solution used for the ASTM G48 standard method for the pitting corrosion resistance testing [2, 4].

Cyclic potentiodynamic polarization tests (ASTM G61) were carried out in a three electrode cell of corrosion measuring system (VoltaLab 10 with VSP unit). Potentiodynamic polarization curves were obtained by the EC-LAB SOFT software. Potential between the sample and the electrolyte has been settled for 10 minutes before the polarization. Scan range was -0.3V – 0.9 V vs the open circuit potential and the scan rate was 1 mV/s, the limit current was 0.15A. The surface of working electrode AISI 304 of 1 cm^2 area was rinsed with ethanol before measurement. The saturated calomel electrode (SCE) was applied as the reference electrode and platinum foil as a counter electrode. Experiments were carried out at ambient temperatures of 22 ± 3 °C and 50 °C. At least five experiment repeats were carried out for all specimens.

Surfaces of working electrode AISI 304 were observed by the optical microscope (Zeis Axio Imager A1m) after potentiodynamic measurements.

Table 2. Conditions of electropolishing.

Component	Content (wt. %)	Temperature (°C)	Current density ($A.cm^{-2}$)	Time (s)
H_3PO_4	64			
H_2SO_4	13	40	0.8	420
H_2O	23			

4. Experiment results and discussion

The pitting potential (E_p) is the main electrochemical characteristic determined from a cyclic potentiodynamic polarization curve, that enables an assessment of the resistance to pitting. After the potential reaches this critical value, current density suddenly increases, denoting the breakdown of the passive film and the beginning of stable pit growth. The shift of E_p to more noble value means the higher thermodynamic stability and the rise of resistance to the pitting [1, 2, 5].

Cyclic potentiodynamic curves of the tested steel in experimental solution at 22 ± 3 °C are presented in figure 3. All three curves have a typical loop shape which points to the breakdown of passive film in the given conditions. According to the common comparison of the curves, the AR surface reflected the most narrow passivity region. The important part of the curves that approximates the interface between the passive state and the active one is enlarged and shown in figure 4 (the location of the respective pitting potentials E_p is indicated by arrows). As can be seen in figure 4, electropolished surfaces have a higher thermodynamic stability and a lower passive state current density than AR one. Fluctuations in the passive current density of AR curve indicate a significantly less stable passive state. Similar differences in passive current density between unpolished and electropolished surfaces describe also authors [11].

The curves of tested steel in experimental solution at 50 °C are shown in figure 5. At this temperature AR curve reflects no passivity state, current density increases immediately after overreaching of its zero value. The surface passive film was probably broken just after immersion into experimental solution. Therefore it was not possible to determine the pitting potential E_p (missing AR curve in figure 6). According to figure 6, there is the strong difference between stability of passive state of EP and G+EP surface. The detail of EP curve shows a slight gradual increase of the passive current density and it points to less stable passive state in comparison with G + EP surface.

The comparisons of the curves (figure 4 and figure 6) and the E_p values (figure 7) point to quality differences between EP and G+EP surfaces. The grinding followed by electropolishing enabled to obtain more resistant oxidic passive layer. This result is consistent with authors [17] who found the key role of grinding prior to the electropolishing in defining the final surface roughness of AISI 304 stainless steel.

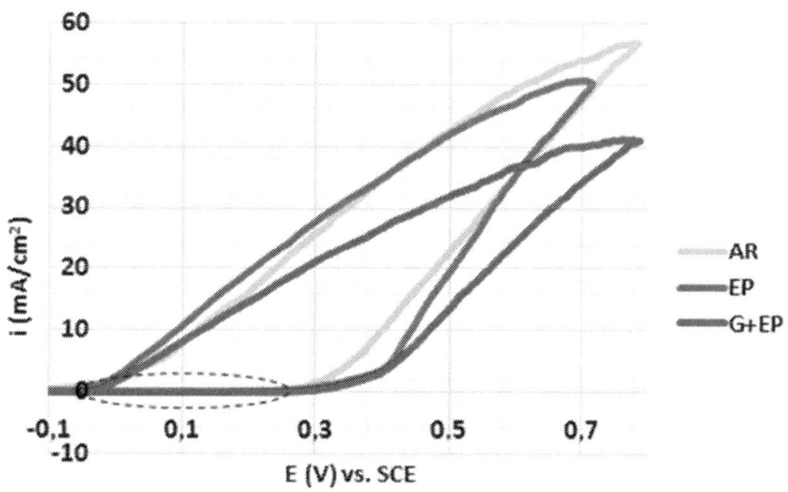

Figure 3. Cyclic potentiodynamic curves of AISI 304 stainless steel in 0.9M NaCl + 0.1M HCl solution, 22 ± 3 °C (AR – as received surface, EP – electropolished surface, G + EP - ground + electropolished surface).

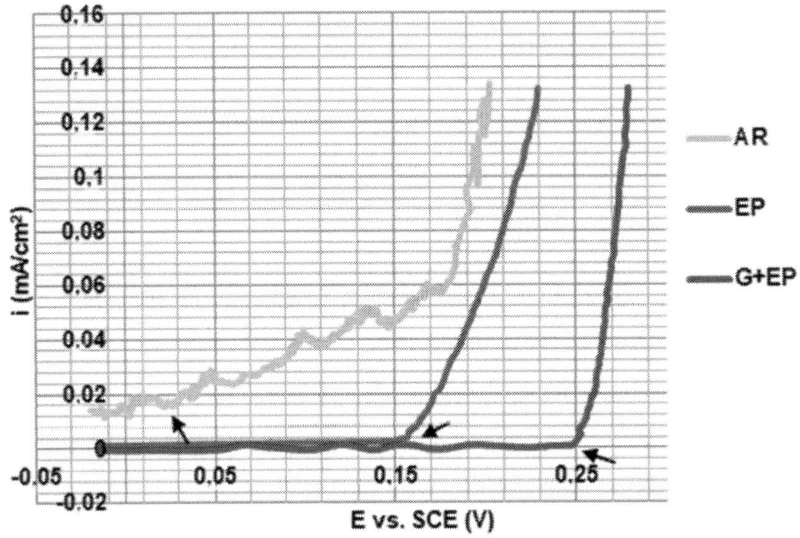

Figure 4. Detail of the cyclic potentiodynamic curves of AISI 304 stainless steel in 0.9M NaCl + 0.1M HCl solution, 22 ± 3 °C (interval 0.0 – 0.25 V; E_p location is indicated by arrows).

According to the comparison of the pitting potentials in figure 7, elevated temperature is the essential external factor that radically decreases the E_p potential even after electropolishing. High quality and chromium enriched passive layer did not prevent adsorption and consequent penetration chloride anions. The passive film was broken at low E_p values and the local pitting corrosion has started.

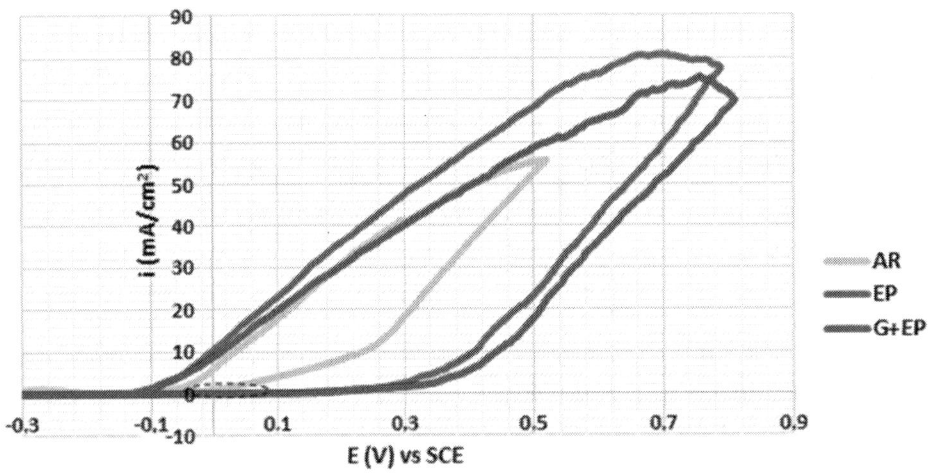

Figure 5. Cyclic potentiodynamic curves of AISI 304 stainless steel in 0.9M NaCl + 0.1M HCl solution, 50 °C (AR – as received surface, EP – electropolished surface, G + EP - ground + electropolished surface).

The setting of the cyclic potentiodynamic test conditions with a sufficiently high limit current (0.15 A) allowed the pitting corrosion progress manifested by well observed pitting. Figures 8 and 9 show pitted surfaces after performed potentiodynamic tests. At the both temperatures AR surface is covered by numerous pits witch are smaller than those on the electropolished surfaces. A worse quality of the passive film could cause its breakdown by chloride anions in many places. It led to the formation of the high number of small pits which size was affected by temperature (figures 8a, 9a;

higher temperature = higher number and smaller size of pits). EP and G+EP surfaces are similar with a smaller number of oval/oblong pits. Chloride anions penetrated through a higher quality passive layer in fewer places, resulting in a smaller number of larger pits (figures 8b, c; 9b, c).

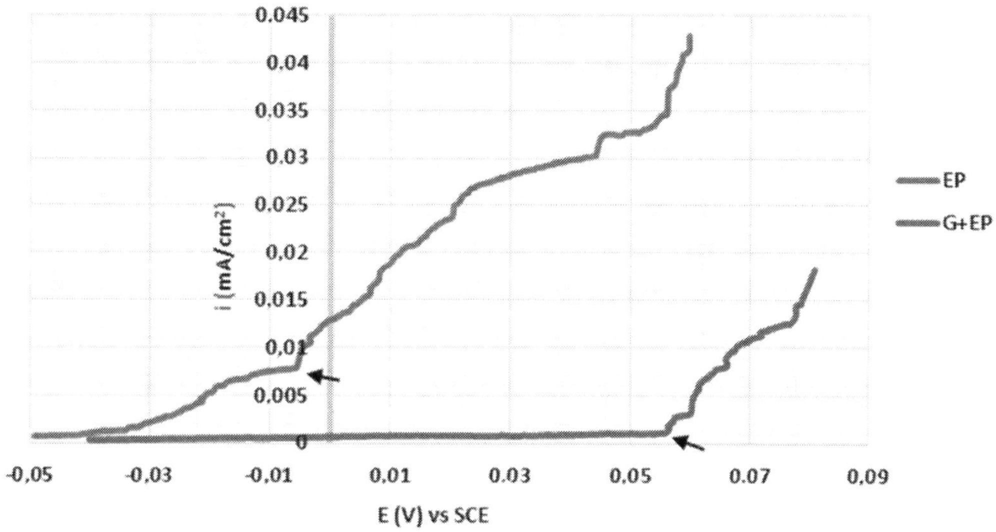

Figure 6. Detail of the cyclic potentiodynamic curves of AISI 304 stainless steel in 0.9M NaCl + 0.1M HCl solution, 50 °C (interval -0.05 – 0.09 V; E_p location is indicated by arrows).

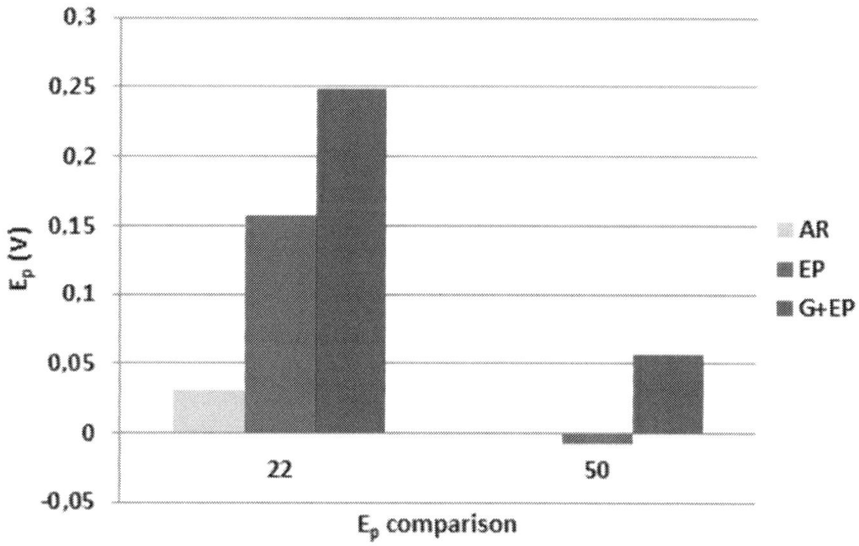

Figure 7. Comparison of the pitting potentials (E_p) at the temperatures 22 ± 3 and 50 °C in dependence on the surface treatment.

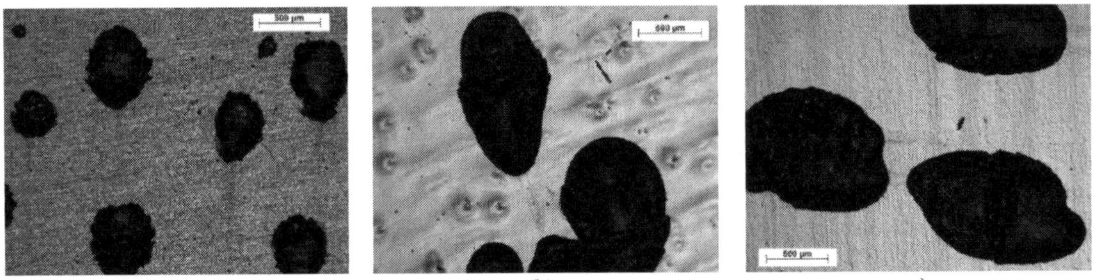

a) b) c)

Figure. 8 Typical pit shape and size on the electropolished AISI 304 steel surface after cyclic potentiodynamic test: 22 ± 3 °C, a) AR surface, b) EP surface, c) G+EP surface.

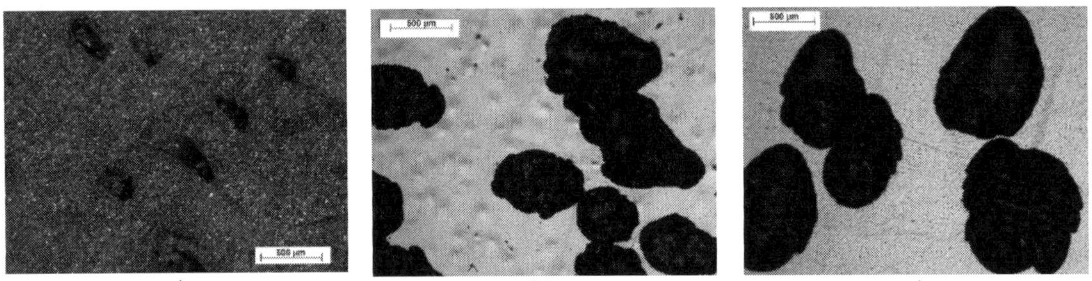

a) b) c)

Figure. 9 Typical pit shape and size on the electropolished AISI 304 steel surface after cyclic potentiodynamic test: 50 °C, a) AR surface, b) EP surface, c) G+EP surface.

5. Conclusion

Based on the results of performed experiments can be concluded:

- The pitting corrosion resistance (expressed by the pitting potentials E_p) of all tested AISI 304 stainless steel surfaces was negatively affected by elevated temperature (50 °C). The most striking negative effect was recorded on "as received surface".
- According to E_p values at both temperatures the ground and electropolished (G + EP) surface showed the most quality passive film and the highest corrosion resistance. However it should be noted that strong E_p decrease at 50 °C points to not sufficient protection against the pitting corrosion in aggressive chloride environment.

In accordance with the results obtained the use of tested steel with the "as received surface" in aggressive chloride environments is not recommended. The ground and consequently electropolished surface seems to be the best alternative. In spite of it caution should be recommended when used in aggressive chloride environments at increased or varying temperatures.

Acknowledgement

The research was supported partially by Scientific Grant Agency of Ministry of Education, Science and Sport of Slovak Republic and Slovak Academy of Sciences grant KEGA No. 049ŽU-4/2017.

References

[1] Szklarska–Smialowska Z 2005 *Pitting and crevice corrosion* NACE International Houston Texas

[2] Liptáková T 2009 *Bodová korózia nehrdzavejúcich ocelí. (Pitting corrosion of stainless steels.)* EDIS Žilina

[3] Uhríčik M, Oravcová M, Palček P, Sapieta M and Chalupová M 2016 *The stress detection and the fatigue lifetime of stainless steel AISI 316L during three-point bending cyclic loading* (Conference Paper) EAN 2016 - 54th International Conference on Experimental Stress Analysis

[4] Zatkalíková V, Markovičová L, Belan J and Liptáková T 2014 *Manufacturing Technology* **14** 3 493-497

[5] Park J O, Matsch S and Böhmi, H 2002 *J. Electrochem. Soc.* **149** 2 B34-B39

[6] Adeli M, Golozar M A and Raeissi K 2010 *Chem Eng Commun* **197** 11 1404-1416

[7] Trépanier Ch and Pelton A R 2004 *Effect of temperature and pH on the corrosion resistance of passivated nitinol and stainless steel* Proceedings of the International Conference on Shape memory and superelastic technologies October 3-7 Baden-Baden Germany

[8] Hong T and Nagumo M 1997 *Corros Sci* **39** 9 1665–1672

[9] Sasaki K and Burstein G T 1996 *Corros Sci* **38** 12 2111–2120

[10] Lee S J and Lai J J 2003 *J Mater Process Tech* **140** 206-210

[11] Momeni M, Esfandiari M and Moayed M H 2012 *Iranian J Mater Sci Eng* **9** 4 34-42

[12] Hashimoto K et al E 2007 *Corros Sci* **49** 1 42-52

[13] Ghanavati S, Shishesaz M R, Farzam M and Danaee I 2015 *Iranian Journal of Oil & Gas Science and Technology* **5** 1 65-72

[14] Available at: http://www.italinox.sk/ (Accessed 14. 6. 2018)

[15] Vaňa D 2014 *Porovnanie technologických možností plazmového a elektrochemického leštenia antikoróznych ocelí v elektrolyte* AlumniPress Trnava

[16] Oravcová M, Palček P, Zatkalíková V, Tański T and Król M 2017 *IOP Conf. Ser.: Mater. Sci. Eng.* **175** 1

[17] Nakar D, Harel D and Hirsch B 2018 Electropolishing effect on roughness metrics of ground stainless steel: a length scale study *IOP Publishing: Surf Topogr Metrol Prop* **6** 015003

Institute of Physics
Dirac House, Temple Back
Bristol BS1 6BE UK

ISSN: 1757-8981
ISBN 978-1-5108-8095-5

9 781510 880955

CURRAN ASSOCIATES INC.
proceedings.com